W9-AHS-200

WITHDRAWN

Inventoried SEP 1976

WILD RIVER

WILD RIVER

Photographs and Text by
Laurence Pringle

J. B. LIPPINCOTT COMPANY
Philadelphia and New York

574.5
PRI

This book is protected by copyright under
terms of the International Copyright Union.
All rights reserved
First edition
Printed in the Netherlands

U.S. Library of Congress Cataloging in Publication Data

Pringle, Laurence P
    Wild river.

    1. Stream ecology—U.S. 2. Wild and scenic
rivers—U.S. I. Title.
QH541.5.S7P75   574.5'2632   78-39756
ISBN-0-397-00811-2

753415

For Alison, and the riffles and pools that lie ahead

Acknowledgments

The publication of this book would not have been possible without the encouragement and advice of nearly a score of people. Its beginnings can be traced back nearly ten years, to the time when Richard K. Winslow helped arrange financing for my first fine photographic equipment. During the 1960s my photographic efforts were encouraged by David Brower, Robert Woodward, Ned Barnard, and Les Line, editor of *Audubon*. An assignment for *Audubon* in 1969 introduced me to several wild rivers in the Adirondack Mountains—the Boreas, Cedar, Goodnow, Indian, and Upper Hudson. It was along these rivers that the idea for this book was born. I am especially grateful to James K. Page, Jr., who helped to further define the concept of the book and guided me to a receptive publisher.

Much of the text is based on studies by generations of biologists and other scientists who have probed the mysteries of flowing water ecosystems. I relied especially on H. B. Noel Hynes's classic work *The Ecology of Running Waters* (University of Toronto Press, 1970). Dr. Hynes and Dr. Kenneth W.

Cummins read portions of the manuscript and suggested ways to improve its accuracy and clarity, as did Jules Tileston of the Bureau of Outdoor Recreation in regard to the last chapter. If any errors remain, they are mine. Alison, my wife, typed and edited the manuscript; it was further improved by the staff of J. B. Lippincott Company, especially by Editor-in-Chief Edward Burlingame.

The photographs cover a span of four years and thousands of miles. Along the way I received photographic advice and assistance from Wendell Dodge, Charles Fontana, John Goodwin, Robin Lewis, and George Mattfeld. Several thousand photos were taken, then gradually narrowed down to the sixty-seven which Joe Crossen used in designing the book.

Many other people offered to help, and I regret that lack of time prevented me from accepting their hospitality and from experiencing more wild rivers. My thanks to all who contributed, directly or tangentially, to the publication of this book.

LAURENCE PRINGLE

Contents

Introduction

River. The word is repeatedly thrust into our consciousness. We read of worsening pollution; we see pictures of a river afire in Cleveland, of dead fish in the Mississippi, Hudson, and Columbia. If we listen carefully, we can hear the sound of political promises breaking. Mercifully, there are things we don't see or hear: factory wastes spewed into rivers under cover of darkness; deals made between industry and government that assure loopholes in pollution legislation.

Rivers bear the brunt of our industrial society. A million years ago, man's impact on river ecosystems was no greater than that of the beaver. Roving bands of people camped beside rivers and paddled these watery roads. They speared fish, dug mussels, and drank the clear water. Wastes thrown into the water were quickly diluted and digested; the rivers flowed sweetly on.

But there are limits to what a river can digest. Today close to four billion people crowd the earth, and many dump sewage and other wastes into its waters. Rivers—the most intensely used of all ecosystems—are further changed by dams, channelization, erosion, biocides, and thermal pollution. Today the sickest rivers are in those nations considered to be the richest, most advanced, most civilized.

Remarkably, even in the United States some rivers still flow clear and wild. They are not untouched; nowhere on earth is there a river unmarked by man's heavy hand. Those who experience their wildness may find beer cans discarded by fishermen, or slash and stumps left by loggers. Even the air may be subtly tainted by pollutants from faraway factories. Nevertheless, these wild rivers are the best we have, and they *are* very fine. Unpolluted, undammed, rich with living organisms, they are priceless laboratories where ecologists can try to unravel a complex skein of aquatic life. Wild rivers also offer unique recreation: whitewater canoeing, rafting, kayaking; tranquil float trips; prime fishing for trout, bass, and salmon. Anyone camping beside a wild river can say, "This is the way it was. This is how the land looked when Indians camped here centuries ago. This is how it looked even before Indians came to the land."

The wild rivers of the United States are national treasures. Some have been preserved in the National Wild and Scenic Rivers System, begun in 1968. Others are being considered for inclusion in this Federal program, but some of them may not survive long enough to gain protection. They are coveted by dam builders and river straighteners, who are amply supported by pork-barrel politicians. Said an administrator of a state water resources development agency, "A good dam site is a valuable natural resource. Most of the best in this country have already been developed. Therefore it is imperative that we acquire as many of the remaining desirable sites as we can."

Considering the burgeoning budgets of the Army Corps of Engineers and the Bureau of Reclamation—the nation's principal dam builders—it seems miraculous that any rivers still run free. By their very success, the dam builders have enhanced the value of the wild rivers that remain. Thus in the battle to protect Arkansas's Cossatot River from the Army Corps of Engineers, conservationists could claim that the Cossatot was made unique by the hand of man when he dammed all the other rivers in the Ouachita Mountains.

This book reveals something of the ecology of wild rivers, their beginnings, their effects on the land around them, and especially the life in and near their flowing waters. The photographs illustrate some of their living and nonliving features. For many readers, these pictures may evoke the sounds, smells, and sights of a wild river valley, offering a vicarious trip downstream on a flowing ribbon of light and life.

This book is also a call to arms. Many of the rivers shown or mentioned have been or are now threatened with a dam or development of some sort. If these rivers are to be saved for the enjoyment of future generations, their saviors will be vigilant, dedicated people who love wild rivers and who express that love by taking effective action.

WILD RIVER

Raindrops on a beech twig reflect the sky, where all rivers begin.

Beginnings, Currents, and Colors

Rivers have quiet beginnings—the whisper of snowflakes on tree boughs, the murmur of raindrops on leaves, the drip of moisture onto forest floors, the trickle of tiny rivulets. With these sounds, a roaring wild river is born.

It can be said that all rivers begin in the sky, with moisture which drops from the clouds in the form of rain or snow. Much happens to the molecules that make up raindrops and snowflakes before they become part of a stream on its way to the sea. In *Living Water,* David Cavagnaro describes these watery adventures:

> What wondrous stories a water molecule could tell, of wild peaks visited on stormy nights, of quiet rivulets and raging rivers traveled, of peaceful fogs and sun-colored clouds, of glaciers and ocean currents, of fragile snowflakes and crisp little frost crystals, and of the seething protoplasmic retorts of living cells—a zillion places visited since the earth's beginning.

Only a tiny fraction of falling rain and snow actually lands in a river or one of its tributaries. Some evaporates and goes back into the atmosphere; some is absorbed by plant roots and returned to the air through transpiration; most seeps deep underground, to emerge later at a spring, pond, or stream or to be stored for centuries. The flow of many rivers in the Western mountains of the United States is sustained almost entirely by meltwater from winter snows, either directly or through groundwater.

Some rivers arise from glaciers: Washington's White River is born at the base of Emmons Glacier on Mount Rainier; the Nisqually River also begins at a Mount Rainier glacier; the Columbia and many other rivers in Northwestern North America are fed in part by glacial streams. Glaciers are formed in places where snow, falling faster than it melts, settles under its own weight and is compacted into ice by succeeding snowfalls. A glacial stream forms from the meltwater at the foot of the glacier and from snow and ice that melt on the glacier's surface. Some of the water in a glacial stream may have fallen as snow hundreds of years ago. At first the water is milky with fine-grained "rock flour" ground from the land by the immense force of the glacial ice, but this material eventually settles out and the stream clears. The discharge of glacial streams is markedly affected by daily and seasonal temperatures; the volume of water is highest in the afternoon and in the late summer, lowest late at night and in the winter.

Sometimes we declare a particular lake the "source" of a river. The Mississippi's official source is Lake Itasca in Minnesota. The Hudson, according to cartographers, begins in a shallow pond called Lake Tear of the Clouds, high on Mount Marcy's shoulder in the Adirondack Mountains. But the wild Upper Hudson really begins in clouds that drop their moisture on the highlands, and in snow and ice that linger on Marcy's peak until June.

Most rivers arise from a mingling of many small streams. Trace these streams backward and you will find even smaller streams, beyond which are little oozing seeps and springs fed by rain and melted snow that have soaked into the soil and then deeper down into porous subsoils and rocks. This groundwater flows and trickles beneath the surface, coming out at low places. The amount of groundwater beneath the land is enormous, thousands of times greater than that which we see flowing in rivers.

In the limestone country of the Southeastern states, full-fledged rivers flow from springs, and many others have springs as major tributaries. One such river is the

*The snow and rain that fall on Mount Marcy eventually trickle
into Lake Tear of the Clouds, official source of the Hudson River.*

Oklawaha of northern Florida, whose flow is augmented by five hundred million gallons of water that gush daily from Silver Springs. Florida's Wacissa River arises from seven major springs which produce sixty-five million gallons daily. In the Missouri Ozarks, the Current River arises from Montauk Springs and is fed by scores of other springs, big and little, along its course. Along the Current and its swift tributary, the Jacks Fork—which were proclaimed the Ozark National Scenic Riverways in 1964, four years before the Wild and Scenic Rivers Act became law—are four springs each of which pours out more than sixty-five million gallons of water a day. One of them, Big Spring, produces a steady two hundred and fifty million gallons daily.

The origin of a river can affect the living things in it. Aquatic insects of a glacier-fed river, at least near the glacier, must adapt their lives to its daily and seasonal fluctuations in volume. Close to their sources, spring-fed rivers provide quite a different environment. The volume of flow is usually steady and the temperature varies remarkably little from season to season. A spring-fed stream seems cool in summer and warm in winter. Its winter "warmth" may enable some aquatic life to survive and to be fairly active, while in another stream the same kinds of creatures are dormant or missing entirely. Because of the uniform temperature of spring waters, they are sometimes homes for animals that cannot survive in nearby streams. One example is a species of minnow that lives only in Kendall Warm Springs, a tributary of the wild Upper Green River in Wyoming.

In the summer, rivers tend to grow warmer as they flow farther from their sources. However, there are exceptions, including the wild Suwannee River of Georgia and Florida, made famous by Stephen Foster's song. Abundant groundwater flows into the Suwannee along its length, and its temperature stays constant for a long distance.

In wild rivers, as in all other ecosystems, living and nonliving factors affect each other in complex and fascinating ways. For example, the flow of a river drops markedly as soon as the trees in its valley leaf out in the spring; the trees send tons of water into the atmosphere through transpiration—water that otherwise might have seeped into the river. The opening of tree leaves along a river also lessens the amount of sunlight reaching the water. Evaporation is reduced and this helps maintain cool water temperatures. The periods of maximum light are spring and fall, and it is in these seasons, rather than full summer, that water plants receive most of the radiant energy they need for photosynthesis. Naturally, the effects of reduced sunlight are greatest on small rivers where the tree canopy shades most or all of the water in the summertime.

Geologists sometimes call rivers "young," "middle-aged," or "mature." These terms refer to the relationship between rivers and the landscape they carve, rather than to a tally of years. One river may "age" rapidly because the land it flows over is easily eroded; another stays "youthful" because its channel is composed of more resistant rock. In the ordinary time sense, rivers are ancient. They began flowing billions of years ago when the earth's atmosphere formed. (Lakes and ponds, by comparison,

Near glaciers and high on mountains, the abundance of stream life is limited by low temperatures in all seasons.

*Canada gooseberry arches over a falls in a brook that rushes
to join others. Many rivers form from a mingling of brooks.*

Once the deciduous trees leaf out, the groundwater available
to rivers and other streams decreases and their flow lessens.

have existed for mere winks of geologic time.) Rivers are not just as old as the hills; they are older. Bit by bit, rivers helped wear away and carry off generations of mountains. And when new mountain ranges reared up, the rivers found new channels to the sea (and sometimes their old ones) and continued to carve away the land.

A mature river flows with gentle curves, and sometimes sweeping meanders, along a broad valley or nearly flat plain. The Wind and Upper Green Rivers of Wyoming and the lower reaches of Washington's Skagit answer this description. Because they are likely places for farms, factories, and cities, mature rivers are especially vulnerable to man. For this reason, many of the rivers that can still be called wild are in the young stage of geological development. They plunge down through narrow valleys, alternately white with cascades and riffles and dark with long pools. Their swift waters are prime targets, however, for dam-builders whose aim is power generation; to them the fast-flowing water is an asset.

The steeper the slope of a river's bed, the faster the speed of its water. Velocity also depends on the roughness of the bed and banks, on the number of bends, and on the various obstructions in the water. The area of swiftest flow is usually beneath the surface, since the surface tension of water exerts a drag. The amount of drag usually increases near the bottom, especially if there is a rough stream bed that offers great resistance to the flowing water. The rate of flow decreases rapidly close to the bottom, reaching zero in a thin boundary along the stream bed.

Most devices used to measure stream velocity do not gauge the flow where the vast majority of plants and animals actually live—on and under rocks, among plants, and just downstream of rocks and other obstructions—but biologists have devised clever ways for measuring the current in these places. Drops of colored water have been released near the surface of rocks and their movement timed over a short distance. Tiny propellers with spinning heads only one centimeter across have been used to measure the rate of flow close to the stream bottom. The German biologist H. Ambühl took photographs to demonstrate that friction slows water velocity near the surface of underwater objects. He released particles of acetyl cellulose in water flowing over a stone and photographed the scene from the side with rapid multiple flashes. The flowing particles appear as long dotted streaks, and the dots of white are closest together near the surface of the stone, where friction slows the particles. The photographs reveal a boundary layer of one to three millimeters where the current slowed almost to a stop, and a zone of still water just downstream of the stone.

Because of friction most of the kinetic energy of a river's rushing water converts to heat energy at the boundary with the stream bed. This causes upward turbulence, which lifts and supports sediments. Turbulence is also caused by sudden changes in slope (such as at the foot of a run), or by a boulder or resistant ledge in the stream bed. The water rebounds off the bottom or the obstacle, producing a standing wave laced with air bubbles. All whitewater, whether in waterfalls or riffles, is the result of air mixing with water.

The roar of a wild river is the sound of water at work, as it tumbles
boulders and pebbles against each other and against the bottom,
wearing it down and grinding the rocks ever smaller.

The sounds of a flowing river are a kind of wild music, but they are also the sounds of water at work, as boulders are worn to cobbles, cobbles to pebbles, pebbles to gravel, gravel to sand, and sand to salt and clay. In the springtime or during any other period of high water, the work pace quickens and the sound becomes a roar punctuated with thumps, booms, and rumbles. Since most of the material in the lower part of the river has come a long way and has been broken up as it traveled, the size of the particles on the bottom decreases as the river flows toward the sea.

A river's surface is marked with mysterious upwellings, puzzling swirls, unexplained eddies. What combination of rocks, flowing water, and physical laws yields such designs? Why has the river laid down a sand and gravel bar in this place? How was the sediment sorted out so that it is coarse on one end and fine on the other? All along the river, mysteries unfold.

When a river's bed is made up of rocks of varying sizes, the water usually flows through a series of alternating riffles and pools. Riffle-pool patterns do not form in streams with bottoms of uniform materials such as sand or silt. Most gravel-bed or boulder-bed rivers have riffles spaced more or less regularly every five to seven stream widths. No one knows why this pattern occurs. Recently some geologists painted exposed gravel in a riffle of a Maryland stream. After a high flow, all the gravel had moved and some pieces were found in the next riffle downstream. The bar itself, however, had the same height and shape as before. Geologist Luna Leopold concluded, "The bar is then a queue in which some particles are at rest, sooner or later to be plucked off the bed and moved downstream, only to be replaced by others."

The gravel, cobbles, and boulders that make up riffles provide a diversity of habitats for aquatic life, and rivers with alternating riffles and pools support a greater variety of plants and animals than do more homogeneous streams. As fishermen know, the turbulence in riffles rips insects free from their moorings, providing food for trout and other fish that lurk below the whitewater. The riffles also provide zones for the mixing of water and air; the water just below a riffle is usually saturated with oxygen day and night. This aeration is vital for the incubation of eggs of such fish as trout and salmon. The rougher the stream bed, the higher and more stable the oxygen content of its water.

Away from riffles the oxygen content usually drops somewhat. At night, aquatic plants stop producing oxygen through photosynthesis and use oxygen in respiration. In the autumn, leaves from trees and shrubs fall into rivers and settle to the bottom in pools and quiet backwaters. There they decay, and the bacterial action consumes oxygen. High waters wash in organic materials, with an accompanying rise in oxygen demand, and also increased turbidity, which reduces the amount of oxygen produced by photosynthesis. The availability of oxygen affects the abundance and distribution of aquatic insects and other life.

The changing colors of a wild river are a result of many factors. At places along the river you may see fluffs of what looks like detergent foam in little stillwater places amid the riffles. Is this pollution in

Sediments are sorted out and neatly arranged by the flowing waters.

Some rivers are colored by mine acids and aniline dyes. The surface
of a wild river reflects the blue of the sky and the gold of aspens.

26

paradise? No. This foam appears in the wildest of streams, whose water is clear and quite safe to drink; it is produced by a combination of chemicals from leaves and other organic matter, mixed and stirred by the tumbling water. It contains great numbers of spores from the fungi and bacteria that are so important in the decay of once-living material.

Fallen leaves and other organic debris also affect the color. Tannins from leaves tint the water varying shades of amber, and in rivers where decay is slow or the load of organic matter is great the water becomes quite dark. But a river's color involves more than its chemistry. Underwater algae and mosses, boulders, logs, and silvery beds of sand also affect its color. Like a moving mirror, the river reflects sky light and the colors of the hills and trees along its shores. In autumn, the river runs red and gold with the reflected fire of maples and aspens. When night falls, it becomes a silvery ribbon in the moonlight or, on dark nights, a black band of mystery.

*The spectacle of a river in autumn diverts us from the ecosystem's
vital processes of energy flow and nutrient cycling.*

Flowing Water, Flowing Energy

Sit by a wild river for an hour and watch some of the workings of an ecosystem. Dragonflies patrol the shore, plucking mosquitoes and small moths from the air. A hatch of mayflies dimples the surface: some are sucked under by trout; others are snatched up by swifts or swallows; still others flutter weakly to shore. A merganser swims by and suddenly dives for a minnow.

You are witnessing strands of the community's food webs, and the interdependence of its living organisms. But some of the most important organisms and processes go unnoticed or may even be hidden from sight. Bacteria, fungi, and algae are growing and reproducing underwater; their presence makes possible the existence of dragonflies and trout. Perhaps you notice

Fallen leaves float for a time, then sink underwater, decay, and become fuel for the river ecosystem.

newly fallen aspen leaves floating downstream and admire the contrast of their gold against the dark waters. For the river, these and other leaves represent gold of a different sort: food energy that helps drive the entire ecosystem.

River ecosystems are strikingly different from those of ponds and lakes, and their ever-flowing current accounts for much of the contrast. Lakes and ponds are receptacles for minerals and other nutrients from the surrounding land, which then cycle and recycle through food webs. In the still waters of these closed ecosystems, plankton are often important food producers. In rivers, the minerals and other nutrients that enter the water are also taken up in food webs, but when they are returned to the water they are carried downstream. They are used again and again, as in lakes, but they move inexorably toward the sea.

Plankton play only a small part in the food production of free-flowing river ecosystems. For many years biologists argued about the existence of plankton in rivers. Running water almost always contains free-floating microorganisms, but most do not qualify as true reproducing plankton. Often the drifting plants and animals flow in from a lake or pond or are washed up from the stream bed. However, true plankton organisms which reproduce in rivers have been found. They are most varied and abundant in big, sluggish streams, far from the swift headwaters. In most wild rivers they contribute little to energy production.

Algae are by far the most abundant plants in wild rivers. They are anchored to all sorts of objects in the water. Some of the smaller species grow in thin films on rocks, providing slippery footing for wading fishermen. Larger species may furnish a substrate to which the tiny varieties cling. Some species occur exclusively in streams; they are adapted to the running water environment that brings continual renewal of gases and nutrients in solution. A wild river offers a wide variety of habitats for algae—riffles or pools, sun or shade, summer or winter, high or low rocks, among other choices. In the riffles, different kinds of algae grow best at different current speeds. In Michigan's Saline River, a biologist found one species of diatom that grew only where water flowed swiftly; it was not found in areas sheltered from the current.

Sometimes great algal populations build up, coating rocks and logs with a brown carpet. Long filaments of green algae trail in the current. When high water scours these plants away, they are not wasted; they become part of the river's food webs. Carried to quiet waters along the shore, to pools, and to still backwaters, the algae settle to the bottom, where animals, bacteria, and fungi begin the recycling process.

Mosses are also a part of a river's plant life, though they are not nearly so plentiful as algae. They thrive near cascades, waterfalls, and rapids, where the water is rich in carbon dioxide as well as in oxygen. They cling to rocks with sturdy rhizoids and flourish in the shade. A thick mat of moss on a boulder may be a home for hundreds of minute animals—rotifers, copepods, scuds, water bears, insects, mites. In quieter waters, you may find cattails, water lilies, and other large rooted plants that

Mosses coat rocks in shady places where carbon dioxide is abundant, and are important sources of food in some headwaters.

Rooted plants contribute food energy to a river's waters in areas
where the current is slow and sediments accumulate on the bottom.

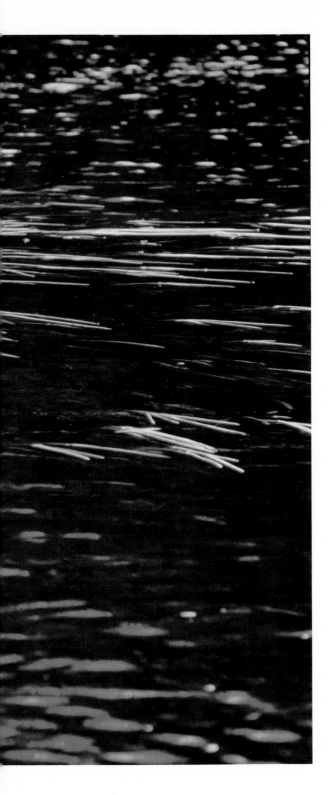

normally grow in ponds and marshes. But most wild rivers have few stillwater refuges for such plants.

Algae, then, are the major food-producing plants in most rivers. Naturally enough, scientists first assumed that algae must also provide most of the primary energy of river ecosystems—the energy that flows through food webs to power the leaps of trout, the flights of mayflies, the dives of otters. However, when they were finally able to measure the primary productivity of rivers, a surprising picture emerged.

A method was devised for estimating the energy production and consumption of water plants, based on the amount of carbon assimilated by photosynthesis. This can be determined by measuring the oxygen content of water as it enters and then leaves a stretch of river. One of several problems with the technique is how to determine the amount of oxygen that enters and leaves the water through the air, rather than from plants. With this factor taken into account, ecologists measured the primary productivity of several rivers and discovered that these rivers often consumed more—sometimes much more—than the amount of energy produced by the algae. It seemed that algae provide only a part—probably a very small one—of the energy in flowing water ecosystems.

Much of a river's living energy comes from the golden aspen leaves mentioned earlier, and from beech leaves, pine needles, fern fronds, and thousands of other sources of organic matter in the river valley. Many insects and other invertebrates feed on this detritus, and especially on the bacteria and fungi that contribute to its breakdown. Some insects feed on water-

By the time a riverside fern begins new growth, nutrients and energy
from its previous year's fronds have become part of insects,
crayfish, trout, and raccoons downstream.

soaked wood. Mussels and the larvae of blackflies are filter feeders; they sift fine particles of leaves and other material from the current. The leaves that enter a river in the autumn are a ready food source throughout the fall and winter. Life in temperate-zone streams does not "shut down" when winter comes; nearly a third of the invertebrates in these streams are active in the winter, and do much of their growing then.

Obviously the trees and other plants of a river valley have a great effect on the living river. Ecologists have noticed that prairie streams are relatively poor in aquatic life, probably because they receive so little organic matter from the plants in their valleys. Headwaters, however, are especially rich in food energy; these streams often flow through forests and are usually narrow, receiving more organic matter per unit area of stream bed than do bigger rivers.

The contribution of energy from the river valley to its waters is not merely in the form of dead leaves. Flowers and fruit of trees, and even the feces of land insects, add to a river's food energy. A river is a kind of year-round "sticky trap" for organic matter from the landscape. In the summer and early autumn, spiders and such land insects as crickets, beetles, ants, inchworms, and grasshoppers either fall, crawl, hop, or are blown into the water, where they become important food for fish. Fly fishermen have known the value of these "terrestrials" for centuries; artificial flies imitating them have proved successful as lures. Grasshopper flies are especially popular with fishermen (and presumably with fish) of the wild rivers in

Tributary streams that flow through forests receive great amounts of dead leaves and other organic matter.

Oregon, Idaho, and Montana. In the late summer millions of live grasshoppers hop about on plants along these rivers, so it seems natural to trout and salmon when a grasshopper fly drops onto the surface.

No one knows the full scope of the dependence of a river's aquatic life on the food energy from its valley. Undoubtedly, it varies from river to river, and from place to place on the same river. A study in Oregon showed that young coho salmon in wooded valleys received more than half of their energy from beyond the river water. And, of course, some aquatic invertebrates are even more dependent on the leaves that fall from the trees.

Facts like these tend to slip from our consciousness when we visit a river in autumn. We hope to be entertained by a flight of geese or by some other dramatic event or spectacular scene, but on some days nothing seems to be happening. No wildlife stirs along the river; the sky is overcast and the foliage past its peak. We head for home, somewhat disappointed, while behind us a silent parade of dead leaves dots the dark surface, and then sinks to the bottom, one by one.

The leaves that enter a river in the autumn are soon broken down and become a major source of food for aquatic life during the fall and winter. Eventually the power of the leaf becomes the power of the trout, the otter, or the osprey.

Life Underwater

The flowing waters of a wild river hold a wealth of life, and the careful observer can sample their richness. Watch as a mayfly appears on the surface. It flutters and skips across the water on delicate pale-yellow wings, then rises and flies toward shore. Another mayfly appears, then a dozen, and soon there are hundreds. A hatch is on. The morning sun glints off scores of wings, and there are splashes and surges in the water as trout and minnows feast on the nymphs still rising underwater and on adults emerging at the surface. As the hatch ends, the frenzy of activity ends too.

Or pick up a stone from beneath the water. The surface may be slick with a fine coating of algae. Turn it over and you will see several odd-looking flattened insects scuttling over the surface, which is dotted with glued-together cases of sand grains or pebbles—the shelters of caddisfly larvae.

The more complex the substrate, the more varied the life in it. A sand bottom supports low numbers of a few species; a substrate made up of rough stones harbors more invertebrates than one covered with smooth stones. The invertebrates that live on, in, or near the bottoms of rivers include leeches, mites, mollusks, sponges, crustaceans, flatworms, and earthworms, as well as tiny organisms such as rotifers, but insects are by far the dominant group. Many insect families occur only in running water, others only in swift water.

Anyone watching water tumble and plunge through a riffle would think it impossible that any tiny insects could exist there—but thousands, perhaps millions, of them do. To survive in the riffle environment, insects must be adapted to live in the thin boundary area close to the substrate or out of the current entirely, under stones or just downstream of them. This they manage very nicely, either through adaptations in behavior or body structure or both.

Many animals that live on stones have flattened bodies which give them a low profile in the current. A striking example is the water penny, the larva of a small land beetle. Its paper-thin oval body is held tightly to stones by its legs and by a ring of movable spines around the margins of its shell. Some species of mayfly larva which live exposed on stones also have extremely flat bodies. However, as H. B. Noel Hynes points out in *The Ecology of Running Waters*, the adaptation of flattening can serve other functions and may occur for reasons other than staying on top of stones: slipping underneath them, for instance.

Beneath the tumbled surface are scores of varied habitats, each the
home of countless microorganisms and of larger creatures
that feed on them.

Most aquatic insect larvae have claws or grapples with which they cling to stones or other rough surfaces. If these devices fail, the animal is swept downstream to almost certain doom. Most of these creatures can't swim, though some species of streamlined mayfly nymphs swim quickly in short bursts. When resting in the current, these mayflies usually stand high on their legs; but where the current is very swift they duck down into the boundary layer.

Many riffle dwellers avoid the areas of strongest current. In large rivers, where the current is much swifter in the middle than along the sides, the bulk of the invertebrate creatures avoid the center. Worms and some insect larvae and nymphs burrow beneath the substrate of the river channel. Each species selects the conditions best for it, whether on top of a stone or in a burrow under it. After an intensive study of Montana streams, a biologist claimed he could predict where he would find eleven species of mayfly on a boulder. In a stream with a surface speed of 240 centimeters per second, three species would be on top of the boulder; two others would be either halfway down the sides or at the top of the downstream side; another species would be anywhere the current was less than 120 centimeters per second; another only where the current was less than 60 centimeters per second; the remaining four species would be under the boulder, two of them living only where detritus had collected.

Current controls the whole structure of a river's living community. As it varies, so does the river's pattern of life. However, the community of plants and animals is also affected by other physical factors, including water temperature, substrate, oxygen content, water chemistry, and periodic changes in the environment such as floods, droughts, and ice. As one's eyes sweep over the details of a wild river, the view shifts from pool to riffle, from sun to shade, from deeps to shallows. Below the surface, just as surely, the communities of plants and animals also vary.

Some of the most remarkable adaptations to flowing water are found among a family of flies called Simuliidae. Because of their humpbacked bodies, they are sometimes called buffalo gnats, but they are usually known as blackflies. Blackflies are only about a tenth of an inch long but have a powerful bite dreaded by Northern fishermen. It is not painful at first, since the insects inject an anesthetic along with an anticoagulant; later, however, the bitten area swells, hurts, and itches. The biological explanation for the biting is simple enough: like mosquitoes, female blackflies require a blood meal in order to produce eggs. For a species to survive, it must reproduce, and the biting females are simply attempting to fulfill their biological destiny. To do so they crawl up pants legs, through buttonholes, and even under shoelaces.

Where blackflies are abundant, they swarm by scores or hundreds around the available food supplies, which are mostly mammals such as deer and moose which cannot hide in burrows as smaller creatures can. Residents in blackfly country stay indoors as much as possible when the adult flies are at their numerical peak, in the spring and early summer. These seasons coincide with prime fishing times, so anglers and others who have to be outdooors have

Underwater, communities of plants and animals vary with the water depth, the speed of the current, and the amount, size, and smoothness of rocks on the river bottom.

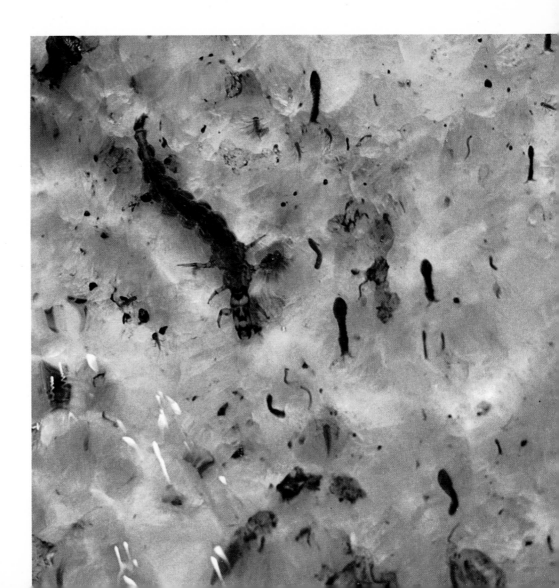

Water flowing in a thin film over quartz reveals several blackfly larvae attached to the rock, a beetle larva preying on them, and tiny stonefly nymphs clinging to niches in the rock surface.

found ways to protect themselves from blackflies; foul-smelling repellents, hats draped with netting to protect face and neck, and clothes of the proper color (khaki instead of blue or red, which seem more attractive to the flies).

Once a female blackfly has her blood meal, the eggs in her body develop further and are ready for laying. Some species fly low over the water and drop the eggs one or a few at a time as they dip in the tip of their abdomens. Most lay their eggs on rocks and logs in swift water at the river's edge. When larvae hatch from the eggs, they move to surfaces that are swept by the fastest current. At times the larvae are so abundant that their massed bodies look like a coating of black moss on the rocks.

Each larva has silk glands with which it spins a tangled mat on its rock. The larva clings to the mat with hooks on one of its two prolegs, which are at or near each end of its body. Normally the larva grips the rock by its rear proleg, and the rest of its body hangs free in the current. Two fanlike organs on the larva's head comb tiny bits of drifting organic matter from the water and sweep them into its mouth. The larva often holds its body so as to bring its food-catching fans out of the slow-current boundary layer that exists close to the rock surface. It can also feed by scraping food from the surface around it.

In order to move, the larva must first make a new silken mat. It bends over so that its head touches the rock surface, spins the new mat, hooks onto it with the front proleg, then lets go of the old mat and bends down again to attach the rear proleg to the new mat. Once attached by its rear proleg, the larva can let go with the front, either to feed or to reach out and spin another new mat. The larva moves slowly, spinning mat after mat, like a mountaineer hammering pitons into the face of a steep rock wall.

The analogy to rock climbing also applies to the larva's safety line. Should the larva lose its grip, it is swept away by the current, but only to the end of a silk line attached to the rock. The larva can then spin a new mat at that point or pull itself up the line to its former position.

After two or three months of eating, the larva spins a cocoon on the rock and transforms into a pupa. The cocoon is open on its downstream end, where the pupa's breathing filaments absorb oxygen from the water. As the pupa transforms to an adult, gas collects under the pupal skin. Enveloped in this gas bubble, the adult emerges from the cocoon, rises quickly to the surface, and immediately takes flight. The adult blackfly may live only a few days—a few weeks at most. Its brief adulthood is devoted to ensuring another generation.

Although, like most aquatic insects, some blackflies have only one generation a year, others produce three or even four generations yearly. Among the many species of Simuliidae, there is also great variation in the timing of egg hatches. Some blackfly eggs hatch soon after they are laid; some last through the winter and hatch when the water warms; others develop slowly over a definite span of months and hatch when the time is up, regardless of the temperature. With such a wealth of adaptations for survival, it is easy to see why blackflies are so successful in the flowing water environment.

Blackflies are an important part of the

*The delicate beauty of mayflies seems to belie this group's
importance in the energy flow of river ecosystems.*

river ecosystem, though it may be difficult to convince a fisherman of this while he is enveloped in a cloud of them. Try later, when his knife slits the gut of a trout and blackfly larvae spill out. The larvae are also eaten by many smaller aquatic predators, including minnows and stonefly nymphs, which are eaten in turn by trout, bass, and other game fish.

To a fly fisherman the most important aquatic insects are the mayflies (order Ephemeroptera). Such artificials as the Quill Gordon, American March Brown, Hendrickson, and Green Drake imitate mayflies. Day in and day out during the late spring, hatching mayflies account for most of the surface activity of trout. (Long before the hatch, of course, mayfly nymphs have nourished trout and other underwater carnivores.) The mayflies of North America are many and varied—about 500 species. Some nymphs burrow in gravel, sand, or the muddy bottoms of pools or of slow-flowing streams. Others scuttle over and under stones in swifter waters. Most nymphs graze on diatoms and other algae growing on rock surfaces. One species has forelegs fringed with hairs which net food from the current. For a long while, biologists assumed that mayflies ate only algae and detritus, but several carnivorous species are now known.

Mayflies are named for the month in which most hatches occur, though the emergence of the adults can extend from March to October. Some species crawl out of the water onto the shore or a stone and transform to adults there, but others simply rise to the surface. Once the nymphs leave the safety of the stones, they are everywhere in the current, defenseless, food for the taking. Fish begin an orgy of feeding.

Floating on the surface, the winged mayflies squirm from their nymphal skins. They often glide on the current for a few moments, supported by the surface tension of the water, or take short, low flights. Fish rise up for them; birds swoop down. At this stage, the mayflies are not yet adults. (Biologists call this subadult a "subimago"; anglers call it a "dun.") The duns rise from the water and flutter to shore. In some species the dun stage lasts a few minutes; in others, a day or two. In all cases the thin, dull dun skin is soon shed and the adult mayfly emerges, glistening and graceful.

In *Downstream*, John Bardach describes the adult mayfly as one of the most elegant and beautiful of all animals. He writes:

When it alights on a twig, standing high on its fragile legs, the upward curve of the abdomen with its sweeping caudal filaments is reminiscent of the sure but delicate brush-strokes of a Chinese painting. The two pairs of wings—the first of which is much larger than the second—are both lacily net-veined; and in color the bodies and wings alike run through almost the entire spectrum of possible hues.

Adult mayflies do not eat. Their mission is to mate. Some live only a few hours, and none more than a few days. The males gather in swarms, rhythmically rising and falling above the water in a sort of mating dance. Once a female mates in the air, she hurries to complete her mission. Some species drop the entire egg mass into the water; others fly low, dipping their abdomens in and depositing a few eggs at a time; still others land on stones, walk underwater, and deposit eggs in protected places there. Some females die just after mating; they fall to the surface and, with a few final twitches, pump eggs out into the water. Other females, exhausted from egg laying, also drop to the surface and float downstream. The fish feed again.

Dead or dying mayflies litter the surface and add significantly to the river's drifting food for a day or two. Year round, however, most drifting organisms are live, immature creatures, not dead adults. Stream biologists first noticed this phenomenon of drift several decades ago, when they caught many insects and other invertebrates in fine-meshed nets set for plankton. A study of one trout stream showed that almost 20 grams (dry weight) of organisms drifted over every square meter of channel in a year. Slower streams carry less drift than swift ones, and riffles produce more than pools. High water increases the numbers of drifting organisms: the greatest amount of drift observed in a Montana trout stream occurred in May, at the time of spring runoff; the least amount in August, during low water.

For yet unexplained reasons, drift increases at night, especially just after sunset. Perhaps the animals become more active in the dark, edging out of their shelters, sometimes to be snatched away by the current. Perhaps they "lose their way" in the dark; some invertebrates orient themselves underwater by noting the fixed objects around them. Nighttime drift decreases when a stream is lit artificially or when the moon is full.

Riffle animals perish if they are carried into a long stretch of slow-moving water. Normally they find a new home in a riffle or are eaten. Fish, especially brown trout, eat many drifters, darting out from shelter to catch them. In fact, this behavior is the basis for the technique of fishing with artificial wet flies.

Drifting invertebrates are known to be an important factor in the flowing water ecosystem, but there are many unanswered questions about the phenomenon. Are the drifters simply "extra" animals that can find no safe place in the environment? What is the relationship between drift and the upstream movements of aquatic animals? Nearly all of them face upstream into the current and apparently move upstream to some extent. When radioactive phosphorus was released in the Sturgeon River of Michigan, it traveled through food chains both upstream and down. In a week, the radioactive substance had spread to snails, stoneflies, and blackflies a hundred yards upstream. After five weeks, the phosphorus was detected five hundred yards upstream in the bodies of stoneflies and fishflies.

Many invertebrates, from crayfish to snails, have been observed to move against the current. Flying insects, especially females ready to lay eggs, also head upstream. If there were no upstream movement, the headwaters of rivers might conceivably lose all their fauna through drift. The complex relationship between drift and upstream movements is one that ecologists have hardly begun to probe.

Stonefly nymphs are an integral part of the life in many wild rivers, though they are much less abundant in drift than mayflies. In Far Northern streams and in the cold headwaters of more Southern ones, stoneflies (order Plecoptera) are usually the dominant part of the fauna. They are adapted to live in very swift currents, and they rarely occur in water where the temperature rises above 77 degrees F.

Stonefly nymphs feed mainly on algae and detritus, though some large and abundant species are carnivorous. They spend anywhere from one to three years underwater. Unlike many mayflies, stonefly nymphs crawl ashore on a rock or other object before transforming to winged flies. In the North some species emerge onto the snow and ice of early spring and sometimes mate and lay eggs. Poor fliers, the adults crawl about more than they fly, but they live longer than mayflies. They feed on the crust of green algae and lichens that grow on tree bark.

After mating, the females fly over the water and drop their egg packets, or touch down on the surface to release a few eggs at a time, or crawl along the water line and deposit them there. The eggs have a jellylike coating which sticks to the substrate underwater. Some small kinds of stoneflies lay eggs in packets held together by a sticky substance which seems to explode as it dissolves, scattering the eggs underwater and probably lessening the competition for food and shelter among the new nymphs.

One order of insects, the Trichoptera, or caddisflies, shows up infrequently in drift, considering the abundance of the larvae in wild rivers, because most of the larvae have evolved extraordinary adaptations with which to succeed in flowing water.

Among the special adaptations of caddisflies are protective cases made of pebbles, sand grains, plant stems, and other building materials from the stream bed. From these simple ingredients the larvae make intricate, beautiful homes—a variety of long and short tubes, slender cones, rectangular stacks of miniature logs, arched cylinders, and dozens more. Some larvae spin delicate silken nets. Certain species can be identified by their distinctive cases. One uses fine grains of sand to make a beautiful replica of the coiled shell of a snail; in fact, the case was first identified as a snail shell.

Close to shore, in quiet backwaters, and in other places where there is little current, caddisfly larvae build cases of plant stems and other lightweight materials, cut to size. The species that live in fast-flowing water use heavier materials; their shelters are less easily swept away. Some species glue large ballast pebbles to the outsides of their cases, making their homes so heavy that swift water has little effect on them; when a larva lets go of the top of a stone where it was feeding, its portable home merely slides off the stone and settles into a cranny. Biologists who have taken caddisfly larvae from their cases and have artificially increased the current around them have found that the insects built new cases of larger, heavier materials.

Case-building larvae begin by spinning a silken tube in which they imbed sticks, sand grains, or whatever building material the species normally uses. Not all caddisfly larvae build portable cases, but they all use silk at one time or another during their lives underwater, if only to spin a cocoon in which to pupate. The case builders are

Cemented sand grains enclose all but the head and legs of a caddisfly larva as it grazes algae from the surface of a stone.

mostly plant eaters, grazing moss and algae from the tops and sides of rocks, or eating detritus from beneath them, or straining food particles from the current with the fringed hairs on their legs. Other larvae spin nets of silk between stones and eat whatever tiny plants and animals the current brings them. Like caddisfly cases, the nets vary in design and location. Some are spun at the brink of miniature waterfalls, yet they seem as delicate as finely woven spider webs. The net spinners hide in part of their net or among nearby rocks.

The case builders are not immune to attack within their shelters; trout relish the larvae enough to gulp them down, cases and all. Nevertheless, the cases offer some protection from predators and sometimes enable the larvae to go undetected among the stones and pebbles on the river bed. This protection has its price, however. Water does not circulate well within the cases,

and the larvae, like nearly all other aquatic insects, must get oxygen from the water around them. The larvae solve this problem by undulating up and down within the cases, which are open at both ends, driving water through them and over their threadlike external gills. The less dissolved oxygen in the water, the more vigorous their movements.

At the end of the larval stage the case dwellers cement their homes to the bottom and seal the openings, and the net spinners fashion rough shelters and spin cocoons inside. Even after transformation to pupae, the insects continue their undulating motions to ensure a supply of oxygen. The pupae resemble adults, with long antennae and legs. At this stage, which lasts only a few weeks, the pupae of certain caddisfly species may become victims of an ichneumonid wasp called *Agriotypus*. The tiny female wasp creeps underwater and finds

47

The dobsonfly lays thousands of eggs on plants overhanging streams. The larvae, called hellgramites, fall into the water as they hatch and are predatory "middlemen" in food webs.

larvae which are about to pupate, or have already done so. She lays an egg on each. Wasp larvae soon hatch and feed on the caddisfly pupae.

Pupae that have not been parasitized rip a slit in their shelters, wriggle out, and swim to the surface or shore. The pupal skin splits and the delicate, rather moth-like adults emerge and fly off. Adult caddis-flies usually emerge at dusk or during the night and are most active at that time, though a few are diurnal; the artificial flies tied by fishermen resemble the latter. They are brown or gray, usually no more than half an inch long, and are easily overlooked in the vegetation at the river's edge. Like mayflies, adult caddisflies live only a short time.

After mating, most females creep into the water and lay their eggs underneath stones, although a few species deposit theirs at the water's edge or on overhanging branches and leaves. Soon after egg laying, the spent bodies of the females float downstream, swirling with the current. Eventually they slip beneath the surface or are sucked under by a trout.

By far the greatest number of aquatic insects are algae scrapers, drift-sifters, or detritus feeders. With some exceptions, the caddisflies, stoneflies, mayflies, and blackflies

A newly emerged dragonfly dries its wings and readies itself to fly from an unfolding fern frond.

obtain most of their food energy from algae, moss, dead leaves, and other detritus. These primary consumers are vital strands in the food webs of wild rivers: they convert plant material to animal tissue. A river ecosystem could retain its stability if its trout or bass disappeared, but not if these aquatic insects died out. The other levels of consumers are far less abundant and important in the total scheme of the ecosystem. Secondary consumers feed mostly on primary consumers. In most wild rivers they include dragonfly nymphs, hellgramites, some stoneflies, and some minnows and other fish. These creatures are "middlemen" in food webs, preying on smaller animals, preyed upon by still larger ones.

It would be difficult to find organisms more entwined in the food webs of a wild river valley than are the dragonflies and damselflies (order Odonata). As underwater nymphs they both prey and are preyed on; as adults they also serve both functions—above the water, along the shore, and sometimes far from the river. The distribution of dragonflies and damselflies, like that of other aquatic invertebrates, is affected by the speed of the current, the water temperature, and the nature of the stream substrate. The nymphs of some species burrow in mud or lurk among plant stems in slow-flowing waters; others are adapted to the gravel bottom and faster flow of the riffle environment. Those nymphs that live in still water or under rocks in fast water rely on their eyes to detect moving prey. This method does not work well for species living out in the current, where turbulence stirs up all sorts of nonfood objects; these nymphs have long legs and antennae which help them locate their food.

When an insect or other small animal comes within range, the double-hinged lower lip or labium of the nymph whips out and impales the prey. At rest, the lip folds beneath the nymph's head, partly covering its face, but it can extend almost half of the nymph's length. It is armed with a pair of jawlike organs, set with teeth or spines to impale the prey. Once the prey is caught, the nymph holds it in front and chews it with strong mandibles. The nymphs of damselflies and dragonflies eat almost any small moving creature, including aquatic and terrestrial insects, crayfish, tadpoles, small fish, and fellow nymphs. As the nymphs grow in size, they select larger prey.

Some species complete their life cycle in a year, but many take longer. In winter a dragonfly nymph hides, eats little, and waits for spring. During its life underwater, a nymph must shed its skin periodically— perhaps a dozen times—to allow for further growth. Finally it is ready for one last change. In the dark of the night, the nymph crawls out onto a rock, plant stem, or some other solid object. Its skin splits down the back, and the dragonfly adult wriggles out. At first it is soft-bodied, with a shriveled abdomen and rumpled wings; but soon blood surges through the wings and they straighten and stiffen. In an hour or two, the squat brown nymph has changed into a beautiful and versatile flying machine.

Adult dragonflies are sometimes called mosquito hawks—an apt name, since mosquitoes are a favorite prey. Both dragonflies and damselflies have bulbous, many-faceted compound eyes. Though no one knows what sort of pictures form in insect brains, biologists assume that many-faceted compound eyes give sharper images than do eyes with few facets. Fast-flying predatory species have thousands of six-sided facets in each eye. The eyes of some dragonflies are made up of 28,000 facets, which probably provide the most acute vision among all insects.

Dragonflies and damselflies scoop prey from the air with a sort of basket formed by their spiny legs. They also pounce on resting insects. The prey may be eaten on the wing or later, while the predator perches. These insects lead an active life through the summer: the dragonflies dart, the damselflies flutter, both pursuing aerial insects and being pursued by birds. But they have only a little while to live; the adults die with the first frosts of autumn. They must mate and ensure another generation. Mating takes place in the air. The paired insects sometimes fly with the male gripping the female's neck with two claspers on the tip of his abdomen. The female dragonfly goes off alone to deposit her eggs, but damselflies hunt for egg-laying sites together. The female's abdomen dips below the water surface again and again, ejecting eggs from her body each time. Some species deposit their eggs below the waterline, on stones or on plant stems.

Once the insects have mated, there seems to be no need for the male to accompany his mate; yet in many species he does. In some cases the male hovers above the female while she dips down and deposits eggs; in others, they fly in tandem, the male attached to and just above the female. Perhaps this unusual behavior enhances the reproductive success of the species. If a swallow or other bird swoops down, it is likely to strike first at the topmost insect of the pair.

Indeed, sometimes a female is seen with part of a male's abdomen still clasped to her neck, the rest of his body apparently snapped off by a bird. After mating he was expendable; the egg-laden female was not.

Crayfish are also important links in a river's food chains. They scavenge dead leaves and other detritus, eat living plants, and occasionally catch and kill aquatic animals. Sometimes they also exploit the food resources above the surface, crawling part way out of the water to pluck at moss on the side of a boulder. In turn, crayfish are eaten by fish, mink, otters, raccoons, and birds. Look along the shore of a wild river and eventually you will find a claw or other crayfish remnant, signifying the passage of some of the crustacean's energy, minerals, and other nutrients to another organism.

A few river creatures hunt the water surface, or just below it. One is *Dolomedes*, the fisher spider, which hides on boulders and waits for insect prey, sometimes pursues it over the surface, and dives beneath the surface after insects and small fish. A silvery coating of air bubbles is trapped in the spider's body hairs, and this oxygen supply enables the spider to stay submerged for half an hour or more.

Water striders (which are true bugs, in the order Hemiptera) also hunt above and just below the surface of the river, though they stay above it all the time. A strider manages to skate and jump on the water by distributing its weight widely on sprawled legs, the tips of which are fringed with waxy water-repellent hairs. The front and back pairs of legs are held stationary while the animal "rows" with the middle pair. Peculiar shadows appear on the river bot-

Energy and minerals from a crayfish exoskeleton are used or released by decomposers and become available for new life.

tom as a strider skims over the sunlit surface; these are cast from the dimples made in the surface by the insect's legs.

A strider's forelegs are shorter than the others and are adapted for grasping prey, mostly small insects. Water striders catch terrestrial insects that fall onto the surface and small crustaceans and aquatic insects that swim close to the surface from below. Most striders avoid turbulent water where the surface tension is broken. One exception is a genus of broad-shouldered water striders with plumes of bristles at the ends of their middle legs. The bristles form a sort of oar which enables the striders to move quickly over turbulent water. The members of this genus can usually be found at the foot of riffles. So far no one knows what they eat, though it is undoubtedly some prey washed to them by the rough water.

Water striders appear to be easy prey for

The water strider's short forelegs grasp small prey from the surface of the water and just below it.

trout and other underwater predators, but they are rarely found in fish stomachs. Perhaps their white undersides serve as camouflage. They also give off odors that are disagreeable to people; this hints that water striders may have evolved chemical repellents that discourage predation.

There are still big gaps in our knowledge of the lives of water striders and their role in river ecosystems. The same can be said of many other aquatic invertebrates, includ-

ing leeches, midges, craneflies, and all the insect orders mentioned earlier. For some groups not even the basic life histories have been worked out, let alone the ecological relationships.

Surprisingly, there is also much to learn about game fish such as salmon, bass, and trout. Certainly they have not been neglected by biologists; indeed, thousands of articles in scientific journals are devoted to the relatively few aquatic vertebrates that provide us with food and sport. But many of the studies treat them only as a resource for man's consumption. They are fodder for fishermen but also play important roles in river communities; we should learn more about *all* their relationships within flowing water ecosystems.

Like the organisms on which they feed, fish are adapted in form, physiology, and behavior to deal with their river environments. Those that swim in swift water have streamlined shapes, round in cross section; those that swim in slow-flowing water have laterally flattened bodies, the "pan fish" shape. Besides these there are fish that live in swift water but seldom swim: darters, sculpins, and some species of minnows and suckers that hug the bottom and hide behind or beneath stones, out of the current. Seldom seen, these fish are indigenous to most wild rivers; one or another species of bottom-dwelling fish lives just about anywhere clear water runs through riffles.

The bottom dwellers are usually flat-bellied, with underslung mouths, eyes placed near the top of their heads, and sturdy pectoral fins with which they prop themselves among stones. Bottom dwellers also have either reduced swim bladders or none at all; these buoyancy-regulating organs are of no use to fish that do not rise to the surface. Bottom-dwelling fish are usually colored to match their stony habitat and can change their color to some extent when the background changes. One species of sculpin, for example, alters its mottled pattern from dark to light in three minutes or less.

Although these fish habitually hug the bottom or hide behind stones, others that swim well also avoid the current whenever possible. Sometimes young trout and salmon spread their pectoral fins in such a way that the current helps them press close to the substrate. Even big trout will seek shelter downstream of boulders, logs, and other obstacles. Fish tire easily; lactic acid builds up much more quickly in their tissues than it does in mammals, and fish take six to twelve times as long to get rid of it. Stream-dwelling fish swim well in short bursts but need to rest fairly often.

In a riffle, a haven of still water is a treasure, something to fight for if necessary. (Riffle-dwelling fish don't defend territories when they are in pools or other quiet waters; in such places a boulder or other obstacle has less value than one in swift currents.) Trout, salmon, sculpins, and darters chase intruders and snap at them. Usually there is no bodily harm; a threat is sufficient. They exhibit this territorial behavior the year round, not just in breeding season. Some species battle only with others of their kind, but many show interspecific territoriality. For example, the Atlantic salmon and brown, rainbow, and brook trout all seem to disregard their genetic differences. In territorial battles, the bigger, healthier fish generally wins—regardless of species—though brown trout (introduced from Europe) often seem to come out on top. Resident fish dominate newcomers, and this accounts in part for the poor survival of hatchery-reared fish that are stocked in streams. A number die of exhaustion after continually being chased from occupied territories.

Some fish begin to defend a territory soon after they hatch, and individuals may spend years in one short stretch of river. In an Indiana creek, a biologist found that rock bass and sunfish had home ranges covering 32 to 63 yards of stream; smallmouth bass and hog suckers covered from 63 to 158 yards. Most of the fish remained in their territories for years, though a few moved far upstream or downstream—for reasons that are still not understood.

The opposite of territorial behavior is schooling. Many minnows and young trout form schools. This gathering of small fish has some survival value. Experiments have shown that predators attack single fish more readily than they attack schools; a school of fish seems to confuse predators. Darkness also protects fish, and many species are most active at night or in twilight. Most fish stay under cover during the day, though the common shiner and some other species are active at that time.

The survival of fish also depends on their physical environment. One of the most vital factors is water temperature, since this determines the dissolved oxygen content and the pace of metabolism in trout and perhaps other species. Oxygen is less soluble and thus less abundant in warm water than in cold water. Some minnows and catfish can survive at temperatures up to 90 degrees F., and carp may live at 100 degrees F.; but most fish need much cooler water. Brook and rainbow trout cannot live long in water temperatures above 77 degrees F. Metabolism rates in trout rise rapidly with the temperature; the same is probably true of other members of the salmon family. Summer temperatures may drive these fish close to the river shore, where the inflow of groundwater and shade from trees cool the water a critical degree or two. Bass, on the other hand, cannot survive in rivers unless the temperature rises to 75 degrees F. or higher in summer. They can live in much colder water—and do in winter—but the summer's maximum temperatures are a vital factor in some facet of the species' survival, perhaps breeding or feeding. The same applies to other species: trout introduced to the cold streams of Glacier National Park do poorly, probably because they are too cold to feed actively in summer. Although bass and trout coexist in some places, their requirements differ enough that waters are often called trout streams or bass streams. In the latter, at least 30 percent of the water area consists of pools or other still water; cold, swift trout waters usually have 15 percent still water or less.

Fish can be labeled as "algae eaters," "detritus eaters," "invertebrate eaters," or "fish eaters." There is, however, some variety in the diet of most individuals, and food also varies with a fish's age, the season, and the locality. Trout feed mostly on drift, and so do several species of minnows. Many suckers and minnows scrape algae from stones; one minnow is called the stone roller because of the way it moves stones around while feeding on algae. Hog suckers also roll stones over, the better to suck up the ooze beneath them.

As a predatory fish grows in size, it takes bigger prey. Black bass proceed from small invertebrates to larger ones, then to crayfish, and finally to an almost exclusive diet of other fish. Fish are often quite selective, consistently passing up one kind of food for another over a period of time; stomachs are often filled with one prey species. Fishermen know of this phenomenon and try to

Trout face upstream, opening their mouths to let water flow over their gills, and feed mostly on invertebrates that drift near.

When trout defend a territory, they are really fighting for an area of stream bed, not the overlying water. The size of the defended territory increases with the size of the fish, but also depends on the availability of large stones and other shelters from the current.

figure out what the game fish are eating on a particular day, so they can offer *their* prey the same kind.

Fish stop eating almost entirely while traveling upstream to spawn; they strike at objects in the water more out of aggression than hunger. Migrants from the sea, probably guided by an internal sun compass, swim three or four thousand miles to the river of their birth. They may journey hundreds of miles farther before reaching the tributary or stretch of river where they were born years earlier. Their incredibly sophisticated sense of smell leads them to the proper place. Many die after spawning, and their bodies supply the river ecosystem with energy and nutrients from the ocean.

Although salmon runs have been greatly reduced by dams, pollution, and overfishing, salmon are still important constituents of wild river ecosystems, especially in the Northwestern United States. The young begin journeying downstream a few months after hatching. They usually drift backward—tail first—so strong are the habits of facing into the current and of getting oxygen by simply opening their mouths and letting water flow over their gills. They turn and swim only if the current is slow or the water is quite warm.

Besides salmon, many other types of river fish also swim upstream to spawn. Some, like sturgeon and shad, come from the ocean, but such freshwater species as trout, suckers, catfish, and some minnows and darters also have spawning runs. Most species of trout breed in the fall, their behavior triggered by the shortening day length. The males swim upriver ahead of the females, moving mostly at night unless the water is high or turbid. Trout spawning

behavior is characteristic of most salmonids. Once the females arrive, they dig pits in the gravel by turning on their sides and flapping their tails so strongly that stones are lifted from the bottom and carried downstream by the current. When the depression in the stream bed, or redd, is several inches deep, the female arches down into it and the male presses close, both right side up. Their bodies vibrate together as eggs and sperm are shed simultaneously. The male swims off a little way and the female moves forward and digs again, sweeping gravel back over the eggs. The trout repeat this process again and again. The fertilized eggs develop in comparative safety under the stones, though any uncovered eggs become food for minnows or other bottom feeders. They hatch after several weeks, and the young trout struggle up through the gravel. They are sustained for several days by their yolk sacs.

Brook trout often dig redds at the bottom of riffles, where water is deflected up through the gravel. Atlantic salmon, steelhead, and brown trout select places at the tops of riffles —the ends of pools—where water flows down into the gravel. In either place, the buried eggs are continually bathed in flowing water. The eggs hatch sooner and the young are bigger where the current is fastest and the oxygen content highest.

Many other kinds of fish build nests in gravel. Some male minnows and chubs carry pebbles in their mouths and pile them over fertilized eggs. Since the piles rise above the substrate, the eggs are bathed in water that flows between the pebbles, thus achieving the effect attained by trout and salmon redds without being restricted to places where water flows into or out of the gravel. These piles are often used as spawning sites by other fish.

Grayling, suckers, many minnows, and several species of darters build no nests at all; in fact, this is probably the most common breeding pattern for river fish. The eggs may be partly covered as the spawning fish writhe on the gravel. Some species have sticky eggs that cling to stones, but most of the fertilized eggs merely settle into niches and develop there. Breeding adaptations include sticking clusters of eggs on the undersides of large stones (sculpins and darters), and laying semibuoyant eggs which develop as they are carried downstream (shad and a few other species). For the great

majority of fish, however, reproductive success depends on a gravel bed, free of silt, bathed in a steady flow of well-oxygenated water.

The lives of fish offer a few spectacles to the untrained eye, and we are usually unaware of the dramatic events taking place underwater. The reflecting surface, the tumbled water, and the shady depths all make it difficult for us to see into their world. While a few adventurers slip below the surface to explore—one biologist, for example, used scuba to observe salmon and trout in a New Brunswick river—most of us glimpse only parts of an intricate whole, and understand much less than we see.

Rivers are linked with the surrounding land and its life in countless ways, many of which are still poorly understood.

Along the River

The relationship between a river and the land around it is one of great intimacy. As biologists learn more about the lives of such creatures as sculpins, crayfish, and mayflies, they discover that these organisms, and thousands of others, are dependent not only on water but on the land and life beyond the river. To some unknown extent, the reverse is also true: many land organisms are tied to river ecosystems and cannot survive without them.

Some dependencies are obvious: a salamander enters the water to lay its eggs; trees flourish along the edge of a prairie river but not a few yards away. Others are more subtle: wildflowers, insects, and birds live north of their normal range because a river's water and valley moderate the climate; spiders that spin orb webs are especially abundant along the shore, as they snare their share of the insects that fly up from the nursery of the river waters. Many kinds of plants and animals live in wild river valleys, but for most it is a hopeless task to assess the degree of their dependence on the river and the life in it. Perhaps it will suffice to describe how a few plants and vertebrate animals are joined with wild rivers, and the rivers with them.

Plants along wild rivers range from cactus to cypress. In the dry "rain shadow" states of the West, water is so precious that the most insignificant brooks have names, and often the only lowland greenery occurs along rivers and their tributaries. In Hells Canyon of the Snake River the summer temperature reaches 116 degrees F. and the slopes are covered mainly with grasses, mullein, sagebrush, bitterbrush, and cactus. Right alongside the river, however, ferns, willows, alders, and cottonwoods grow, nourished by its waters. The river is a long oasis, and the shoreline vegetation it supports provides most of the food energy for life in the water.

In the East and other areas where rainfall is abundant, forests often grow to the river's edge. Most of the trees that overhang the water are no different from those in the surrounding woods. The dark waters of the Suwannee are shaded by cypress, ash, gum, magnolia, and live oak, their branches hung with Spanish moss. The Allagash is sheathed in spruce-fir forest; the Upper Hudson, in a deciduous hardwood forest of yellow birch, beech, and maple, interspersed with cedar, pine, hemlock, and other conifers; the Current and Eleven Point, in a mixed hardwood forest of maple, basswood, tulip tree, hickory, oak, and walnut.

Often a river flows past a variety of

A fritillary named "aphrodite" alights on a thistle alongside the Allagash.

All along wild rivers, spiders spin orb webs and catch numerous flying insects. A fallen leaf signals the arrival of autumn and the end of a spider's life.

Deciduous forests with
scattered conifers
clothe the worn mountains
that enclose the Upper
Hudson and many other
Northeastern
wild rivers.

plant types on its journey through the land. As the Klamath River cuts through the Siskiyou Mountains and on to the Pacific, it passes three distinct forest types in the span of about a hundred miles. In the upper reaches the principal trees are ponderosa pine, Douglas fir, sugar pine, Jeffrey pine, and red and white fir; in the middle reaches, they are Douglas fir, oak, western hemlock, and madrone; as the river approaches the ocean, it flows past redwood, bay, western hemlock, and red cedar.

Some trees thrive only in wet bottomlands along the river. They include many species of willow, alder, and ash, as well as swamp maple, sycamore, and tamarack. Beneath them in the moist soil live shrubs, ferns, mosses, and wildflowers such as jack-in-the-pulpit. The very edge of the river, and sometimes the river channel itself, offers a sunlit though challenging environment for plants. Swept by floods, battered by ice, subject to sudden rises in water, the shore nonetheless is often lush with ferns, grasses, wild roses, asters, and wild iris. Cardinal flowers, columbines, and bluets sprout from little pockets of soil on riverside boulders, and annuals of various kinds take their chances on transitory sand and gravel bars.

The wet bottomlands of forested river valleys are rich with
mosses, wildflowers such as wood sorrel, and mushrooms
such as the deadly fly amanita.

A single late blossom stands out among the bunchberries that cover the forest floor.

The abundant fruit of bunchberries brightens the river bottomland forest, while cardinal flowers thrive among boulders along the sunlit shore.

The boundary area between two ecosystems is called an ecotone. Life is varied and abundant wherever these areas occur, a fact well known to bird watchers and hunters. The river's shore is an ecotone between the river and the forest, prairie, or whatever ecosystem dominates the river valley. This narrow ribbon of land is visited by nearly all the large creatures of the adjoining land ecosystem, and some make it their special province. Likewise, some animals of the river ecosystem spend most of their time, or critical periods of it, in the ecotone.

A springtime chorus of trills, twangs, cheeps, and chirps announces the mating season of frogs and toads along the river. Frogs live in and along all wild rivers; even in a valley of torrential rapids they find refuges of still water. One primitive species, the tailed frog, lives in cold headwater streams of the Northwest. Its strings of eggs are attached to the undersides of

Seeds are deposited on sand and gravel bars. They may sprout and grow, but their survival depends on the whims of the river. Frogs are creatures of the river-land ecotone, eating many terrestrial insects and seldom venturing into deep water or fast currents.

stones, and its tadpoles have a large sucking disk which holds them to objects in fast water. Some frogs live in the river the year round, snapping their share of mayflies and other insects from the air. Others, such as wood and tree frogs, come to the river only to mate, then scatter back to the surrounding land. The same is true of most toads, which may travel considerable distances to reach a river pool or other quiet backwater in which to leave strings of fertilized eggs. Along the Middle Snake, Rogue, Colorado, or other rivers which flow through arid lands, toads tend to stay close to the river and its tributaries all year.

While toads and frogs fill the river night with rhythmic calls, salamanders mate silently. Some are entirely terrestrial, though their eggs must be laid and must develop in damp places under logs and stones; others spend years of larval life in the river and as adults live on nearby land; still others live only in rivers. Turn over a rock along the river edge and a salamander will dart from beneath it, looking for another dark, wet place. These amphibians are largely nocturnal. Most crawl about on the river bottom, feeding on insects and other invertebrates, though the hellbender, a large salamander which may grow to be as much as twenty inches long, eats crayfish and small fish along with its usual scavenging diet. Most salamanders are too small or too few to loom large in a river's energy relations. With a few exceptions, amphibians are still-water creatures and select the least riverlike portions of running waters; the same is true of most reptiles.

Turtles and snakes are usually the only river reptiles, though alligators still live in some rivers of the Atlantic and Gulf Coast states. Most turtles are omnivores, eating both plants and animals. Their life cycles are the reverse of those of the salamanders who live on the land but lay their eggs in the water; the turtles spend most of their lives in the river and go ashore only to lay eggs. There are no turtles at all in some Western wild rivers. The Suwannee River, on the other hand, has a turtle named specifically for it—the plant-eating Suwannee turtle of northern Florida. Most turtles are more common in ponds and lakes than they are in rivers, but usually a few can be seen basking on logs in stillwater stretches. Soft-shelled turtles, a true riverine group, are common in rivers east of the Rockies. The main aquatic snakes of North America are the water snakes, which are found over a widespread area, and the cottonmouth, or water moccasin, which occurs in the Southeastern states and as far west as Texas and southern Illinois. Many terrestrial snakes hunt the river edges, especially in dry seasons or arid country. They are all predators, catching large insects, frogs, toads, and mice.

There are two main sources of sound along wild rivers—the flowing water itself and the birds that are drawn to it. The river offers quite a variety of sounds, from musical trickles to rhythmic roars, but nothing to match that of the birds. In the springtime, their dawn chorus fills the river valley with music. Most of the songs are inspired by territorial urges, but there are also calls of alarm, assembly, and food getting. The notes of one particular bird might be considered the theme song of wild

A dipper emerges from beneath the waters of the Salmon River, holding a minnow in its beak.

rivers. This bird is the dipper or water ouzel, a species that lives along the wild, rushing streams and rivers of the Western mountains. Its song, always accompanied by the sounds of flowing water, is a rich mixture of trills and flutelike notes; it has a joyous lilt. Gray and starling-sized, the dipper is as much a creature of the river as is the tailed frog of the Northwest. It seems to fear land and builds its mossy, ovenlike nest close to the water, often right beside or even behind a waterfall.

At first sight, the dipper seems much like another river-edge bird, the spotted sandpiper. Both fly low over the water, hop from rock to rock, and probe underwater with their bills. But the sandpiper, a common shorebird across the entire continent, teeters its tail up and down as it walks, while the dipper bobs its entire body up and down. The sandpiper catches small insects and crustaceans in shallow water; only rarely does it step below the surface in pursuit of prey or to escape danger. The dipper, however, feeds underwater much of the time. Caddisfly larvae are a favorite food. In shallow water it steps along head down, busily turning this way and that after insects, with only its back showing. In deeper water it "flies" to the bottom and walks about. Returning to the surface, the bird may pause to preen, oiling its feathers for the next plunge. In winter the dipper continues much as before, except to fly downstream to warmer altitudes if the water in the highlands freezes over. To some wilderness enthusiasts, the hardy little dipper embodies the essence of a river's flowing wildness.

Another avian predator, more widespread than the spotted sandpiper, is the belted kingfisher. It feeds higher on food chains than the sandpiper or dipper, taking mostly minnows, frogs, and crayfish. Flashing down from a streamside perch, the kingfisher dives underwater and strikes at a fish, selected from above, with its stout beak. Instantly it rises and returns to the same lookout place or to another farther along the river. If no perch is handy, the bird may hover over the water, then plunge in again. Once a minnow or other prey animal is caught, the kingfisher perches on a limb, pounds the prey against the wood to stun it, then gulps it down head first.

The rattling call of the kingfisher echoes along the river as the bird patrols its territory. In spring its sturdy bill is used to excavate a deep tunnel in a bank of soil along or near the river. Often the kingfisher's nest is just one of dozens in such a bank, the rest housing bank swallows. While the swallows hunt insects over the river to feed their young, the paired king-

fishers bring hundreds of minnows to their half-dozen offspring. The kingfisher takes some young trout and salmon too, but not so many that it affects the numbers of legal-sized fish available to anglers. A river has shelter and food for only a certain number of good-sized fish; one way or another many young must die so that a few may live, grow to maturity, and reproduce.

The great blue heron also helps cull fish populations, though its diet is more varied than the kingfisher's. This great bird, which is four feet high and has a wing span of six feet, feeds in water where it can wade and in meadows along the river. Its sharp beak snatches up fish, frogs, salamanders, crayfish, eels, and snakes. On land it catches mice, ground squirrels, and gophers, sometimes taking them to the water and dunking them to wet their fur before swallowing. In water it stands still, its rapier-like bill poised above the water, or moves along with careful steps, its long neck and head stretched ahead of its feet.

Usually a great blue heron is a picture of deliberate grace. Only rarely is this image shattered, as when a heron stabs a big fish and is dragged underwater for a time or loses its balance. Fishing in a swift-flowing river, one heron, in striking at a fish that darted between its legs, lost its balance and floated downstream on its back, still clutching the fish in its beak.

The great blue heron nests in a treetop colony in a secluded swamp or on an island. Each tree may support several bulky nests made of sticks, twigs, and grasses. If the colony is undisturbed, the nests are used year after year. The adults range for several miles around the colony in search of

Arriving at its nesting tree, a great blue heron
alights with silent grace.

food. When a parent returns, it drops lightly onto the nest and regurgitates a partially digested meal into the gaping bills of its offspring. The young grow to adult size in a few weeks and clamber about on nearby branches, testing their wings. Once free of the nest, herons wander in all directions before starting their southern migration. In the autumn, great numbers gather in their favorite hunting areas, such as the John Day River in northern Oregon. The great blue heron is a hardy bird, and some individuals spend winter in the North, even in areas where the temperature drops below zero. Like the dipper, the great blue heron can survive if enough water stays free of ice.

Green herons and snowy egrets sometimes patrol the river edge, though like most other wading birds they favor the still waters of marshes and ponds. Bald eagles eat dead fish washed onto the shore, including spent salmon from the great wild rivers of the Northwest. They also swoop down and pluck live fish from the water. If the catch is too heavy, the eagle rows ashore, using its wings as oars.

The survival of the American bald eagle is threatened by illegal hunting, by habitat destruction, and especially by persistent biocides that now exist to some degree in every body of water on earth. Chances for the preservation of this majestic bird will improve when long-lasting biocides are banned from the continent and when its habitat of wild rivers, lakes, and coastal bays is saved from further destruction.

A bald eagle sometimes bullies an osprey into dropping its prey, then eats the fish itself. Ospreys fish in the same manner as kingfishers, though their plunges into the water are more spectacular, as befits a bird with a five-foot wing span. Diving from a hundred or more feet in the air, an osprey strikes feet first, not head first, and seizes its slippery prey with strong talons. Rising from the water with powerful wing strokes, it carries its catch to a dead tree overlooking the river and rips it into bite-sized chunks. It hunts over fairly clear waters where it can see the fish that swim near the surface. The numbers of osprey have been reduced by the same factors affecting bald eagles.

Many other birds of prey hunt in wild river valleys, including the anhinga of the South, but most are not so directly dependent on water. Golden eagles and Swainson's hawks soar over the rugged countryside traversed by the Snake, Salmon, Grand Ronde, and many other Western rivers. Prairie falcons hunt the same dry, open lands, snatching meadowlarks and blackbirds from the air but also feeding on the superabundant grasshoppers. Goshawks prey on red squirrels, mice, and grouse in the forests along the Allagash, Upper Hudson, St. Croix, Wolf, and other wild rivers of the North. In Florida, the rare and spectacularly graceful swallow-tailed kite is sometimes seen along rivers, plucking a dragonfly from the air or swooping low over the surface to sip some water. At night, the comments of a flowing river are often punctuated by the calls of barred and great horned owls.

Waterfowl of many kinds nest, feed, and rest along wild rivers, especially in still-water stretches, nearby sloughs, and beaver ponds on tributary streams. Migration brings flocks of geese and ducks to the

Wild rivers are refuges for Canada geese and other waterfowl during migration. The rivers also serve as overwintering places when ponds and other still waters freeze over.

rivers, where they pause for days or even weeks, feed and rest, then fly on. Thousands of migrating ducks and geese stop along the Snake, downstream from where it is joined by Oregon's Powder River. During August and September, large numbers of goldeneyes gather in the upper reaches of the Yellowstone, and thousands of snow geese spend the winter in the delta area of the Skagit.

Mergansers nest along good-sized clear-water streams; they dive after fish, frogs, insects, and crayfish, which they pursue by sight. When ice covers the surface of the long pools and other still waters, they fly to swift waters. Many dabbling ducks, including mallards, teal, pintails, and wood ducks, nest along wild rivers and their bottomlands. Trumpeter swans grace the Madison, Yellowstone, and Lewis Rivers of Yellowstone National Park and vicinity.

Birds like these catch our eye and help form our image of a river's important wildlife. Actually, all birds play a rather small part in a river's energy budget and, depending on the season and the river, the significance of herons, eagles, and waterfowl in the energy flow may be less than that of other smaller birds in the river valley—the warblers that flit through the

Morning light along a river's edge reveals tracks left by nocturnal mammals—a raccoon, and perhaps a mink.

trees and shrubs; the swifts that seize insects above the water and nest in hollow trees; the swallows that vie for their share of the insect bounty; the sparrows, thrushes, and wrens that nest and feed along the river edge and add their music to the sound of flowing water.

At night the insect patrol is taken up by bats. Often their abundance is determined more by the availability of shelter than by the amount of food obtainable. If there are hollow trees, caves, rock crevices, or abandoned mines fairly close to a river, bats will flit after insects in the night sky over the water and nearby open areas. The most spectacular display occurs in New Mexico, where a half million free-tailed bats pour out of Carlsbad Caverns at dusk and scatter to hunt over the Pecos River valley. Along other rivers the bats are fewer and often go unnoticed.

Other small mammals catch river insects beneath the surface. Along Eastern rivers and their tributaries, star-nosed moles dive down and eat aquatic insects and crustaceans, locating them with sensitive fleshy feelers on their snouts. The Northern water shrew is even more aquatic. Fringes of bristles on its hind feet and tail enable it to swim and dive well. With air bubbles clinging to its fine-haired pelt as it enters the water, the water shrew looks like a miniature silver submarine. Since the bubbles give it buoyancy, the little creature holds onto objects on the bottom as it hunts for insects, tadpoles, and other small prey, then lets go and bobs to the surface. Once ashore, it shakes itself dry like a dog. Like all shrews, it has a tremendous metabolic rate, with a heartbeat of nearly eight hundred times a minute, and must search for food almost constantly. Although the water shrew occurs along many rivers and streams of the North and of the Southern mountains, its small size (about six inches, including its tail) and nocturnal habits make it one of the least known mammals in wild river valleys.

In contrast, we easily associate raccoons with a river shore. These omnivores leave their childlike paw prints along river edges from Florida to Maine and from British Columbia to Mexico. They are particularly abundant in wooded bottomlands where hollow den trees and logs are available. From the river, raccoons harvest frogs, crayfish, salamanders, insects, mussels, and fish; an even more varied diet, including nuts, berries, turtle eggs, bird eggs, and nestlings, comes from the land. Raccoons of wilderness areas are identical with those that raid suburban garbage pails.

Some mammals have more of an aura of wild country about them. The elusive mink is one. Its hunting is concentrated in the river-land ecotone, where this dark, graceful weasel investigates every hollow log, rock crevice, or other hideout for potential prey. An expert swimmer and diver, it catches a good share of its food underwater. In the early summer, a pair of minks must range far from the den in order to feed themselves and their young, usually about six in number. Often the parents stockpile food while the young are nursing, preparing for the time when their offspring stop mouthing mouse carcasses and frog legs and begin tearing into this food in earnest. Life becomes less frantic for the parents when the family hunts as a group.

This is the time you are most likely to see them, loping along the shore on a summer evening, pausing to probe in a burrow, racing on to the next hideout, wheeling and dashing into the water after a frog. The mink family moves along the riverbank like a sinuous, flowing brown line.

A favorite prey of the mink is the muskrat. A healthy adult muskrat acquits itself well against an attacking mink, but many young, sick, injured, or otherwise handicapped muskrats are easier prey. The muskrat is usually thought of as a mammal of ponds and marshes, and it is most abundant in such places. However, it also lives in rivers, sometimes even in wildly turbulent ones. It feeds on mussels and shoreline plants. Sometimes it builds a house of mud, cattails, and other plant materials in quiet backwaters; more often along rivers, however, it digs a bank burrow with an underwater entrance that slopes upward to a plant-lined chamber used as a shelter and a nursery. The muskrat swims and dives with ease, forepaws tucked into its chest, webbed hind paws powering it through the water. Its scaly tail is laterally flattened and seems to serve as a rudder. As the muskrat swims, only its head and a bit of its back and tail stick out of the water. Sometimes it is difficult to tell a swimming muskrat from its much larger relative, the beaver, though a beaver has a squarer snout.

Hike along almost any wild river in the United States and sooner or later you will find some light-colored sticks, stripped of their bark, lying along the shore, a sign that a beaver colony lives somewhere upstream, either on the river or on one of its tribu-

taries. The ways of beavers are familiar to most people—their dam building, their diet of tree bark, their loud tail splashes that warn of danger. However, each one of these characteristics can be countered by others, equally true: some beavers don't build dams; beavers prefer soft plant foods to tree bark; the tail splash signal probably has additional meanings.

Most wild rivers are too broad and brawny to be tamed by a beaver dam; but this does not restrict beavers to the tributaries, for the main functions of a dam are simply to provide a barrier of water—like a moat—around the lodge, and a refuge from which the beavers venture to get food. The same security can be found along a river, with a burrow dug in the bank and a stretch of the river serving as the pond. Sometimes the bank den is difficult to find; however, beavers often build a partial lodge over it, and if the den is occupied in the fall, the lodge is repaired with branches and fresh mud and is easy to spot along the riverbank.

In the fall, a bank den can also be detected by the presence of branches and small trees piled near the underwater entrance. This is the beaver colony's winter food supply. Bark and twigs are the only available winter food for beavers over much of their range. In other seasons, beavers choose softer foods that are easier to digest: grasses, ferns, mushrooms, algae, and the leaves, stems, and roots of such water plants as cattail. As long as these foods are available, beavers cut few trees unless they are needed for building or repairing dams or lodges.

Many tributary streams of wild rivers

are of a size that beavers can dam. Often additional dams are constructed, above and below the first one, enabling the colony to exploit new food sources in comparative safety. Once the dams are built they require little maintenance except in periods of heavy rain or melting snow.

The dam-building behavior of beavers was studied recently by Swedish biologist Lars Wilsson, who worked with the European species. He raised one beaver for a year and a half, all the while making sure it neither saw nor heard running water; then he put a loudspeaker at one side of a tub of water in the beaver's cage and played a recording of running water. The beaver began piling sticks into a sort of dam at that side of the tub.

Dr. Wilsson observed other beavers repairing a dam. He made a deep furrow in the top of the dam so that water poured through, then put a clear plastic hood over the furrow and stuffed bits of wood around it so he, and apparently the beavers, couldn't hear the sound of water rushing underneath. He made another furrow a short distance away, above the waterline. When the beavers came to the dam, they

Beavers eat bark and twigs when they must, but prefer softer foods. A colony usually consists of an adult pair, a few yearlings, and the young of the year (kits).

quickly filled the second (dry) furrow with sticks and mud. This didn't stop the leak, of course, but it seemed to satisfy the beavers. They failed to find the first furrow because its plastic cover was even with the top of the dam and they could not hear the water rushing out. When the plastic hood was replaced with an opaque net that did not muffle the sound, the beavers quickly covered the net with bits of wood until the running water could no longer be heard.

Later, Dr. Wilsson discovered that he could make the beavers begin work on the dam simply by playing the sound of running water. Even though there was no leak, they added material to the dam at the source of the sound. From these studies it appears that beavers make dam repairs wherever they hear water rushing or see an unevenness in the top of the dam.

Dr. Wilsson's beavers not only began building to the sound of running water, but also responded to the sound of an electric

Tall grasses soon cover the mud of an abandoned beaver pond,
and the forest slowly encroaches on the meadow, closing in
on the stream which carries away pond sediments from its old course.

Wild rivers and their tributaries are havens for the river otter,
a species that delights and thrills many people, and that depends
on the availability of wild lands and waters for its survival.

razor! Apparently they react to oscillating sounds of certain frequencies. Dr. Wilsson is trying to define the stimuli more precisely. Perhaps the frequency of the sound even affects what sort of material a beaver brings to repair a leaking dam. North American biologists are now checking their suspicions that Dr. Wilsson's findings apply to beaver on this continent. It appears that the building behavior of beavers is controlled by inborn reactions, a conclusion that is no surprise to ethologists, but may disappoint those who like to think of beavers as shaggy civil engineers without hard hats.

As beavers cut, dig, and build, their activities affect many other organisms. In the span of a few months after a dam has been erected, a portion of the wooded valley is changed to an open pond; the water keeps air from reaching tree roots, and the trees die. Some organisms thrive in the new environment, others perish, and the species composition changes as the pond ages and accumulates silt, leaves, and other organic material. The invertebrates of running water are replaced by stillwater species and, for a time at least, trout prosper. However, once the beavers have exhausted the supply of winter food in the area, they move on. The pond may persist for several years, but eventually the untended dam breaks open and most of the pond waters drain away. A carpet of grass soon covers the mud. Deer, bears, grouse, and many songbirds are attracted to these beaver meadows, where the stream flows along its old course, past standing dead trees.

A beaver pond is often only a few acres in size. However, the watershed of a wild river may contain scores of such ponds and beaver meadows in various stages of succession back to forest. Thousands of acres may be affected as beavers shape the environment to fit their needs. The effects on the wild river valley are profound.

If the dipper is the avian symbol of wild rivers, its mammalian counterpart is the river otter. Words like "graceful" and "supple" take on new dimensions when one sees a river otter at play or at work in the water. Backflipping, diving, sliding on mud or snow, gliding on its belly down a wild cataract, the otter exhibits extraordinary strength and muscular coordination.

Maps in field guides show otter range covering all but the Southwestern desert regions of the United States. Regrettably, these maps are nearly half a century out of date. Otters have been wiped out in several states and are rare, near extinction, in many others. They are holding their own only in areas where rivers still run clear and wild. Fishermen have had a part in the otter's decline; believing that otters eat too many game fish, some fishermen kill otters illegally in the hope that this will improve the fishing. Their suspicions about otters are supported by the writings of Izaac Walton and Ernest Thompson Seton, who, unhampered by facts, declared that otters ate great numbers of trout, salmon, and other game fish. However, in the past thirty-five years there have been dozens of investigations of otter food habits, based on examinations of droppings or of stomach contents, and they all refute the idea that otters are serious competition for anglers.

Otters eat many things, from bass to blueberries; their food varies somewhat with the season and locale. Crayfish are a favorite food, often making up a third of the diet. Fish may make up another third, but they are mostly such creatures as minnows, suckers, darters, sculpins, mad toms, and sticklebacks. Otters eat the fish that are most abundant and easiest to catch. These are usually not game species; in fact, many of them—such as darters and sculpins—are animals that sometimes eat the eggs of trout. A careful investigation of the otter's role in nature thus would probably reveal that it is actually beneficial to the fish that anglers value. However, it is long past time to stop justifying a creature's existence on the basis of its value to humans. As this anthropocentrism declines, the otter's chances for survival in North

The howls of wolves, which can still be heard along some Northern wild rivers, add immeasurably to the aesthetic quality of wilderness.

Autumn finds deer moving down to the refuge of river bottomlands.

America will increase. In his book *The World of the Otter,* biologist Ed Park concludes:

> Even if the otter were found to feed almost exclusively on trout, many of us would gladly let it have its share just for the thrill of occasionally seeing this fur-clad torpedo, bubbling with happy spirits, cruising a wooded shoreline or sliding joyfully down a snowbank. Fishes are so very plentiful; otters so very rare.

Wild river valleys are refuges for other large mammals, including such endangered species as wolves and mountain lions. In the autumn, deer and elk move down from the hills and mountains to the milder climate of the valleys. Moose also winter in the bottomlands, feeding on willows and aspens along the shore and sometimes on succulent plants in the water. They can be seen along the Allagash, Big Fork, Salmon, and Clearwater Rivers, as well as near the Snake, Madison, and other wild rivers in the area of Yellowstone and Grand Teton National Parks and along the many wild rivers in Alaska. Black bears live in the same areas, as well as much farther south, near rivers that arise in or bisect mountain ranges, and also in Florida. They sometimes prowl river edges in search of berries, eat dead fish or those they catch themselves, and rip apart logs in beaver meadows, looking for insects and rodents. Grizzly and Alaskan brown bears gorge on salmon during spawning runs, wading into the shallows of riffles and grabbing fish with their jaws or flipping them ashore with their paws.

Some mammals and birds are infrequent visitors of the river edge; others are utterly dependent on the river's waters. But all figure in the flow of energy and materials. Together with the mayflies and diatoms, their great numbers and diversity make the ecosystem whole. A wild river is more than water flowing in a channel, carving the land, depositing sand bars where herons leave tracks, supplying oxygen to fish and insects, sustaining shoreline willows. Its influence reaches much farther, to elk that summer in the mountains, to geese that fly to the Arctic, and to salmon in the seas.

The harmful effects of channelization, dams, and other stream
"improvement" projects reach far beyond a river's immediate channel.

Wild Rivers and Man

There are more than three million miles of rivers and tributaries in the United States, and very few of them appear as they did two or three centuries ago. Rivers were then jumping-off places for exploration; now they are dumping-off places for wastes. Rivers have been altered and dammed for flood control, navigation, hydroelectric power, water supply, and irrigation. Some of the changes were needed and well thought out, but most were not. Pollution and the other ecological insults to this nation's rivers have already been well documented in books and in leading environmental magazines. Every year there is new evidence of harmful side effects, but each year brings new proposals for mindless tampering with flowing-water ecosystems.

During the past four decades, the Army Corps of Engineers, the Bureau of Reclamation, and especially the Soil Conservation Service have been "improving" streams through channelization. The usual treatment involves clearing all vegetation a distance of one hundred feet from the bank,

deepening the stream channel, and eliminating all natural meanderings in order to speed the flow of high waters, thus reducing local flood damage, and to drain adjacent wetlands for agriculture. To some extent, these purposes have been achieved (though at great expense). Only in the past few years, however, have the damaging side effects become known. Consider Missouri's Blackwater River, for example. Part of the river was channelized about sixty years ago. As a result, local flooding lessened and flood-plain land was freed for farming: so far, so good. But the channelization led to increased flood damage farther downstream and also caused serious erosion along the banks of the river and its tributaries. The river channel is growing wider and wider, and most bridges along its course have had to be replaced or lengthened several times. Similar disastrous side effects of channelization have been reported from North Carolina and Louisiana. Unfortunately, results like these are sometimes used to justify further harmful activity by the agency that started it all.

Channelization also causes significant losses in wildlife populations, either through drainage of flood-plain wetlands or destruction of stream habitat itself. In a study of twenty-three North Carolina streams, biologists found that the numbers of game fish had been reduced by 90 percent as a result of channelization. According to a 1970 study by the Tennessee Game and Fish Commission, a proposed channelization project in that state's Obion and Forked Deer river basins would result in an annual loss of nearly three million dollars in fish, wildlife, and recreation values. The project was halted, at least temporarily, when conservationists convinced a court that the Corps of Engineers' project violated several laws, including the National Environmental Policy Act. Although environmental impact statements are required on all Federal projects of this sort, many of these reports are incredibly inadequate. The Soil Conservation Service statement on the environmental effects of twenty-six miles of "channel improvement" on Georgia's Chattooga River, for instance, contained only this sentence: "No adverse effects on man's environment are anticipated."

Early in 1971, national opposition to channelization finally achieved some results: the Soil Conservation Service temporarily halted all projects and ordered its chief officer in each state to review existing plans and make sure that channelization was carried out with minimum loss of fish and wildlife. However, the order did not allow the alternative of leaving a stream in its natural condition.

The Soil Conservation Service has already "improved" thousands of miles of streams while completing 284 projects. Over

Decades of dam-building and river channelization have made
unspoiled, free-flowing rivers increasingly rare and valuable.

a thousand new projects have been approved by the SCS, applications are pending for three thousand more, and the SCS estimates that eleven thousand other streams "need" to be channeled. Understandably, conservationists are not pleased with this prospect.

Many of the existing or proposed channelization projects involve parts of rivers and streams that flow through farmland and near cities and industries. Some of the streams are polluted. However, they all support some fish and wildlife and provide outdoor recreation. Some are truly wild. In Georgia, the SCS projects completed, approved, or proposed include thirteen streams suggested by that state's Natural Area Council for designation as State Scenic Rivers. And channelization has already destroyed much of southern Florida's Kissimee River, changing a wild river into an ugly ditch and draining life-giving water from an estimated quarter million acres of marsh.

In the arid Southwest, another form of stream destruction is being proposed and practiced, though only to a limited extent so far. The goal is to provide more water for irrigation; the method is to remove all deep-rooted trees and shrubs from riverbanks. These riparian plants soak up water, most of which transpires from their leaves into the air. In Arizona the plants involved are mainly mesquite, salt cedar, and—in higher altitudes—cottonwood. Studies conducted on the Gila River by the United States Geological Survey showed that an average of two acre-feet of water seems to be saved annually by removing riparian vegetation. This figure is enticing to government irrigation agencies and to the few growers who would reap vast profits from these tax-supported projects. The Corps of Engineers already has authorization to remove river-edge plants along most of the length of the Gila.

Riparian plants have extraordinary value to wildlife in arid country. In Arizona, such animals as deer, javelina, doves, quail, gray hawks, and black-bellied tree ducks would be adversely affected if riverbanks were stripped. When Arizona biologists compared the fish population in a portion of a stream where the banks were cleared with other stretches having uncleared banks, they found 80 percent fewer fish in the stripped portion. Opposition to river-edge stripping projects is mounting among biologists, anglers, and others who see ecological folly in the making.

Human activities beyond the immediate river valley may also seriously affect the living stream. Several outstanding wild rivers flow through national forests, where timber harvesting has recently been increased and where overgrazing is a long-established practice. Clear cutting, in which every tree of every age is felled in a given section of forest, is the common timber harvest method. Experiments in New Hampshire showed that clear cutting a watershed results in greater stream flow, erosion, and loss of nutrients from the soil. Careless logging methods can choke a stream with mud and silt, and the effects of such assaults on the environment can reach far downstream. In Idaho's Payette and Boise National Forests, greatly increased erosion as a result of logging has virtually wiped out the great salmon and steelhead spawning grounds of the South

Water and the riparian plants it sustains are especially beneficial to wildlife in arid country, such as along the John Day River in Oregon.

Fork of the Salmon River.

Many a good trout stream in the East has been ruined by the effects of nearby strip mining, as silt washes in and covers spawning beds and as acids from the mines kill all living things outright. Pennsylvania's Clarion River is marred by acid mine drainage. In the 1968 Wild and Scenic Rivers Act, the Clarion was designated as a potential addition to the national system. However, it is doubtful that the Clarion will meet the guidelines used to evaluate free-flowing rivers, specifically that "the river should be of high quality water or susceptible to restoration to that condition."

Strip mining also threatens the quality of the Big South Fork of the Cumberland

River, in Kentucky and Tennessee. This is one of the most spectacular rivers in the East, nestled in a deep valley enclosed by massive bluffs. There are, however, nearly eleven thousand acres of strip-mined land in the river's basin, and two to three times more acreage there may be stripped in the next twenty-five years.

Of all the harm man does to flowing rivers, none is worse than a dam, which completely destroys a living river and its valley, sometimes for a distance of fifty miles or more. The resulting dam lake bears little resemblance to a natural body of water. Usually the water level rises and falls greatly over the span of a year, and this prevents the development of a normal and varied lake-edge environment. The lake becomes a sink for sediments which will eventually fill it, though at a faster pace than most natural lakes are filled. Often the outflowing water comes from great depths, and its year-round cold temperature reduces the variety and abundance of life for some distance downstream. In addition the water may be chemically unusual, containing ferrous, manganous, and sulphide ions, along with great quantities of bacteria and dead algae. This also affects downstream life adversely.

Dams block salmon and other fish on their spawning runs. Even when fish ladders are provided and some fish manage to pass one dam—and the next, and the next—to spawn, their young have great difficulty getting downstream to the sea. They follow the bottom contour of a river, not the surface, and when they reach the upstream side of a dam, they mill about near the base; they have lost their way,

Gliding down a wild river is not unlike visiting an art museum.
Masterpieces of nature appear in the distance and nearby.

Trees cling to a cliff face above the river. Along the shore are collections of water-worn stones that represent millions of years of earth history.

and many die. Young and adults are also killed by nitrogen gas, which is caught from the air and concentrated as excess water falls over dams. Under great pressure in the basins below dams, the nitrogen dissolves into a supersaturation of the gas and blinds, cripples, or kills fish swimming there. This "gas bubble disease" is presently killing salmon by the millions below dams on the Lower Snake, the Columbia, and other Northwestern rivers.

Fortunately, there are signs that the free-wheeling days of dam building are near an end. As we learn more about their disastrous side effects, their failure to provide promised benefits, and the statistical tricks used to justify them in the first place, it seems likely that fewer and fewer dams will be built. In fact, as more of this sort of evidence is revealed and we become more ecologically sophisticated, it is quite possible that some dams will be torn down. Many conservationists believe that this task would give the Bureau of Reclamation its first useful work in many years.

The Bureau and the Army Corps of Engineers attempt to justify projects through an economic yardstick called the "benefit-cost ratio." The construction and operation costs of a project are weighed against various benefits such as flood control, water supply, power production, irrigation, recreation, and fish and wildlife enhancement. Congress doesn't usually approve funds for projects with unfavorable ratios, so the dam-building agencies go to great lengths to "reduce" costs and dream up benefits. One example: for a dam on Arkansas's Cossatot River, the Corps claimed a benefit from the assumption that industries would be attracted to the area downriver from the dam. Then the Corps claimed yet another benefit: water from the reservoir could be used to dilute pollution caused by these same industries.

Of the variety of tricks used to exaggerate benefits, the greatest deceptions occur on the cost side of the ledger. Until recently the interest charges on construction funds were figured on low rates that hadn't been available for decades. In 1972, the Water Resources Council proposed new guidelines with more realistic interest costs. However, the costs of lost wildlife habitat, harm to fisheries, and other side effects of dams and their lakes continue to be underestimated or excluded entirely. All this would not be so infuriating if account juggling had not been the key difference in decisions on many water projects through the years. Congress relies on the benefit-cost ratio as a statistical summary of a project's relative desirability. Yet a high-ranking Bureau economist said recently, "Congressmen unfortunately take the benefit-cost ratios seriously. Really they are not too meaningful."

Unfortunately, Congress is far from innocent in this regard. When pork-barrel politics demands it, Congress has been known to send the Bureau or the Corps back to recalculate and somehow create a favorable benefit-cost ratio for some boondoggle. One of the most recent, accurate, and thorough indictments of government folly in river valleys is *Damming the West,* the Nader Task Force Report on the Bureau of Reclamation. Published late in 1971, this well-documented book describes in detail the Bureau's unneeded irrigation projects, the damage to the environment, the benefit-cost distortions, the hidden subsidies to profiteers, and the infringements on Indian water rights. Some of the findings can, no doubt, be applied to the Corps of Engineers and its flood-control and navigation projects. Ideally, it should take just one copy of *Damming the West* presented to each Senator and Congressman to cause a drastic reduction in the activities of the dam-building agencies. Realistically, it will take a great public outcry to achieve this effect and to halt the further waste of money and river valleys.

The behavior of conservationists perplexes some people who say, "Conservationists are always negative, always trying to stop something. You can't halt progress." A wide

Living and nonliving objects are arranged in unique
compositions by a river's water.

Whitewater offers a variety of thrills and challenges, ranging from the rushing wildness of a kayak ride through huge standing waves to a gently bumpy raft float through a riffle.

gulf in understanding exists between conservationists and their detractors. A word like "progress," for example, can be applied to endeavors other than highway building and river damming. The late Howard Zahniser of the Wilderness Society wrote:

Out of the wilderness has come the substance of our culture, and with a living wilderness—it is our faith—we shall have also a vibrant, vital culture, an enduring civilization of healthful happy people who like Antaeus perpetually renew themselves in contact with the earth. . . . Working to preserve in perpetuity is a great inspiration. We are not fighting a rear-guard action, we are facing a frontier. We are not slowing down a force that inevitably will destroy all the wilderness there is. We are generating another force, never to be wholly spent, that, renewed generation after generation, will be always effective in preserving wilderness. We are not fighting progress. We are making it.

As for the negative image of conservationists, one of their foremost spokesmen, David Brower, has said, "If you are against something, you are for something. If you are against a dam, you are for a river."

When a river is tamed by a dam, a variety of unique recreation values are lost. Water skiing is no substitute for whitewater canoeing, and the experience of a fisherman casting flies for trout on a wild river is quite different from that of fishing on a wide, still reservoir.

The popularity of whitewater sports has grown tremendously in recent years, and

Olympic competition in these sports, begun in 1972, will spur further interest. The Hudson River Whitewater Derby, started in 1958, now attracts thousands of spectators and hundreds of contestants. It is held near North Creek, New York, in early May, when the Upper Hudson is at its roaring best. The races include giant slaloms, with as many as twenty gates, for canoes, kayaks, and covered canoes, and a 7½-mile downriver race. Contests of this sort are held on several wild rivers when the water is high. But whitewater is also run in every other season (except winter in the northernmost states), when there are no watching crowds, just people in fragile crafts pitting their skills against the rocks and water. It is no place for a novice; a beginner should stick with an experienced river runner, preferably one with iron nerves, quick reflexes, and an ability to read the water.

Whitewater is deceptive, appearing faster than it really is. Chances are it is flowing only ten miles or, at its wildest and whitest, twenty miles per hour. But it seems fast enough, perhaps too fast, as one kneels in a canoe, gliding down those last few smooth yards, listening to the rising roar and seeing spray and waves just ahead. Even at ten miles an hour, the water exerts great force; in this current a seventeen-foot canoe would be pressed against a boulder with a force of seven tons. The remains of crumpled and torn canoes and kayaks are sometimes seen along whitewater stretches—another reminder that beginners need not apply.

Wild rivers offer many other boating experiences, including peaceful floats in johnboats and canoes on Arkansas's Buffalo River and wild—but safe—rides in dories or rubber rafts on the Colorado River through the Grand Canyon. Commercial raft voyages are now available on the Middle Snake through Hells Canyon, on the Middle Fork of the Salmon, the Rogue, the Rio Grande in Texas, and on several other rivers, including the Stanislaus, American, and Klamath of California. Where the Snake flows through Grand Teton National Park, the float trips are sometimes too popular, with rafts queuing on busy days. At other times, a raft floats alone on the curving river and its occupants are likely to see moose, beaver, herons, and eagles. Some of these rivers can also be traveled in a jet boat, a quick means of transportation to a fishing spot or camp but hardly a craft for truly experiencing a river's wildness. Sometimes people ride down rivers in inner tubes from truck tires or go tubing through a stretch of whitewater, then walk up the shore and float down again.

Many raft and canoe trips last several days. The craft crunches ashore on a gravel beach in the late afternoon, and camp is set up. There is time to explore a side canyon, where the campers find a fresh bear track or an Indian petroglyph. Then the night closes in around the spellbinding magic of a campfire's coals. In the morning, mist hides the opposite shore and the river looks like the edge of the world as it disappears into a gray void.

Backpackers and trail riders travel the river valleys along paths carved in rock faces by Indians or early settlers. They may follow in the footsteps of explorers such as Lewis and Clark, who traveled on or near a number of Northwestern rivers. Long stretches of their route are now under the

Children gather around a morning fire to warm themselves and wait for the sun to burn away the mist that rises from the river.

The glow of a campfire's coals awakens something deep within us, and evokes a feeling of kinship with the Indians who camped beside the same river thousands of years ago.

waters of dam lakes, but other parts are not; some of the expedition's campsites can still be located along the Upper Missouri in Montana, on the last major undammed stretch of this river. Deep in Hells Canyon of the Snake River is Dug Bar, where in 1877 Chief Joseph led the Nez Percé across the river—then in full flood—as they made a valiant effort to reach refuge in Canada after being driven from their ancestral lands by the U.S. Army. Then there is the Rogue River Trail, originally blazed by gold prospectors in the 1880s. Further back in time, river valleys were the homes of this continent's first peoples. Hundreds of archaeological sites have been discovered in wild river valleys, and many more remain to be found and studied.

The nation's finest stream fishing is found along wild rivers. The Buffalo and St. Croix Rivers are famous for their smallmouth bass; the Madison River of Montana and Wyoming is considered one of the best trout streams in the West. Undammed rivers running into the Pacific Ocean offer extraordinary fishing. All five species of Pacific salmon run up Washington's Skagit, where steelhead fishing is especially fine. On Oregon's Rogue River, fishing for steelhead and salmon is also outstanding during the seven months of the year when spawning runs are on. California's Klamath River is the world's largest producer of steelhead.

Wild river valleys are rich with wildlife, and hunters find plentiful game such as deer, elk, moose, and wild turkey. For many game species, the value of the river extends far beyond its immediate valley. White-tailed deer travel a dozen miles or more to winter in the valley of the Upper Hudson. Hells Canyon of the Snake River provides critical winter range for thousands of deer and elk. In normal winters, about twenty-eight thousand mule deer and five thousand elk winter below the four-thousand-foot level of the Snake River canyon and that of the Imnaha River which joins it. In severe winters they move deeper into the valleys. Without these refuges, game populations would be reduced for many miles around.

To some extent, defenders of wild rivers can play statistical games with dam-building agencies and cite the number of visitor days that would be lost if a dam were built. A value can be assigned to deer and elk herds and the figure plugged into benefit-cost studies. But some values defy quantification. There are people who never expect to visit a wild river and yet are pleased that such rivers exist. Perhaps their children or grandchildren might float or fish a wild river someday. As Wallace Stegner has written, "We simply need that wild country available to us, even if we never do more than drive to its edge and look in. For it can be a means of reassuring ourselves of our sanity as creatures, a part of the geography of hope."

Luna Leopold has attempted to quantify some of the aesthetic qualities of river valleys, working out a system in which evaluation numbers are assigned to physical factors such as river width, pattern, and degree of bank erosion; biologic and water quality factors such as turbidity, fauna, flora, pollution evidence, and diversity; and human use and interest factors such as vistas,

One of the greatest values of wild rivers is their solitude.
With each child born, each highway extended, and each transistor radio
manufactured, the opportunities to be alone decrease,
and the preciousness of wild rivers increases.

historic features, and the amount of trash and litter. Of the human interest features, Leopold writes:

> These are often more intangible than either the physical or the biological ones, but they are nevertheless influential in determining how the landscape impresses us. For example, if one is at the point on the Delaware River where George Washington is supposed to have thrown the silver dollar, that historical incident, however apocryphal, gives that place a distinct meaning.

When Leopold applied the forty-six criteria of his evaluation system to twelve river valleys in central Idaho, the Snake River in Hells Canyon scored highest in positive uniqueness. Leopold also devised two scales for evaluating "river character" and "valley character" and applied them to the same twelve rivers; the Snake in Hells Canyon ranked first again, with the Salmon River at Carey's Falls close behind. Finally, he compared the Snake in Hells Canyon with four rivers already preserved in national parks—the Merced in Yosemite, the Colorado in Grand Canyon, the Snake in Grand Teton, and the Yellowstone in Yellowstone National Park. Once more the Snake in Hells Canyon ranked high, second only to the Colorado in the Grand Canyon. Leopold concluded, "This analysis shows that the Hells Canyon of the Snake River, the site proposed for a hydroelectric dam, has unique characteristics that give it an exceptional esthetic rating and place it in a category shared by few other landscapes within the United States."

The 1960s were a decade of great environmental awakening in the United States, as people began to sense the extent of the damage done by the mindless misuse of technology and to realize how little truly wild land was left. After a long struggle, the National Wilderness Protection Act of 1964 was passed. In the course of that struggle, conservationists wondered what else should be saved, before it was too late. The answer, of course, was wild rivers. Urged on by conservationists, the U.S. Departments of Agriculture and Interior compiled a list of 650 rivers, or parts of rivers, which seemed to have special qualities. Sixty-seven of these were investigated briefly, twenty-two in greater detail, and the studies, completed in 1964, became the basis for legislation. In October, 1968, the Wild and Scenic Rivers Act became law, beginning with these words:

It is hereby declared to be the policy of the United States that certain selected rivers of the Nation which, with their immediate environments, possess outstandingly remarkable scenic, recreational, geologic, fish and wildlife, historic, cultural, or other similar values, shall be preserved in free-flowing condition, and that they and their immediate environments shall be protected for the benefit and enjoyment of present and future generations. The Congress declares that the established national policy of dam and other construction at appropriate sections of the rivers of the United States needs to be complemented by a policy that would preserve other selected rivers or sections thereof in their free-flowing condition to protect the water quality of such rivers and to fulfill other vital national conservation purposes.

The Act established the Wild and Scenic Rivers System. The eight initial rivers to be protected were: the Rio Grande in New Mexico, the St. Croix in Minnesota and Wisconsin, and the Wolf in Wisconsin, to be administered by the Interior Department; the Eleven Point in Missouri; the Middle Fork of the Feather in California, and the Middle Fork of the Clearwater and the Middle Fork of the Salmon in Idaho, to be administered by the Department of Agriculture; and the Rogue in Oregon, to be administered jointly by both departments. These rivers secured early protection because they were noncontroversial, being mostly on Federal lands, of undisputed national importance, and not part of any dam plans by a government agency.

Twenty-seven other rivers were named in the Act to be studied for possible inclusion in the wild river system. The studies, to be completed over a ten-year period, are being made by Federal-state teams led by the Bureau of Outdoor Recreation (representing the Interior Department) or the Forest Service (representing the Department of Agriculture). The rivers being studied by the Bureau of Outdoor Recreation are: the Allegheny, the Clarion, and Pine Creek in Pennsylvania; the Bruneau in Idaho; the Buffalo and the Obed in Tennessee; the Delaware in Pennsylvania and New York; the Gasconade in Missouri; the Little Beaver and the Little Miami in Ohio; the Maumee in Ohio and Indiana; the Missouri in Montana; the Penobscot in Maine; the Rio Grande in Texas; the St. Croix in Minnesota and Wisconsin (a stretch of the river south of the part already preserved); the Suwannee in Georgia and Florida; the Upper Iowa in Iowa; and the Youghiogheny in Maryland and Pennsylvania.

Many "old swimmin' holes" now have high coliform bacteria counts and sunken shopping carts, but one can plunge into a deep pool of a wild river without qualms.

Many anglers prefer stream fishing to lake fishing, yet each year the miles of available streams are reduced by dams, or their productivity is severely harmed by channelization.

The Forest Service is responsible for studies of these rivers: the Chattooga in Georgia, North Carolina, and South Carolina; the Flathead in Montana; the Illinois in Oregon; the Pere Marquette in Michigan; the Priest, the Moyie, the Saint Joe, and the Salmon in Idaho; and the Skagit in Washington.

These twenty-seven rivers were not included in the national system because they are controversial; there was strong support to keep them free-flowing, but some were coveted by dam builders and many others flow through private and state as well as Federal lands. However, all these rivers are protected for several years from dams, power lines, channelization, and any other Federally financed developments while studies continue and legislation on their fate is considered.

Although the 1968 Act allowed ten years for completion of all the studies, the current schedule calls for completion by 1976. Meanwhile, forty-seven other rivers in twenty-four states await their fate. These rivers were selected under section 5(d) of the Act that called for the Secretaries of Agriculture and Interior to name further potential additions to the national system. Unlike the twenty-seven listed above, these rivers are given no protection by the Act and will not even be studied until work is complete on some of the initial group. However, if an agency such as the Corps of Engineers begins planning a project on one of these rivers, the agency is required to investigate the values of the free-flowing river along with those of development. This guarantees that wild river status is considered along with other potential uses.

During the preliminary studies of the early 1960s, it became clear that there were beautiful free-flowing rivers and parts of rivers that were worthy of protection but that were not wild in the wilderness sense. Some rivers, like the Middle Fork of the Salmon and the upper portion of the Suwannee in Okefanokee Swamp, are true wilderness rivers. But others, such as the Upper Iowa, flow through agricultural land, and many have roads and even low dams along their courses. To allow for these degrees of wildness, the act provides for the inclusion of three classes of rivers in the national system:

"Wild river areas" are "vestiges of primitive America." They are "rivers or sections of rivers that are free of impoundments and generally inaccessible except by trail, with watersheds or shorelines essentially primitive and waters unpolluted."

"Scenic river areas" should also be "free of impoundments, with shorelines or watersheds still largely primitive and shorelines largely undeveloped, but accessible in places by roads."

"Recreational river areas . . . are readily accessible by road or railroad, . . . may have some development along their shorelines, and . . . may have undergone some impoundment or diversion in the past."

All but one mile of the Middle Fork of the Salmon River is considered a wild river area, and the Allagash, added to the national system in 1970, is administered as totally wild. However, most of the rivers that are now in the system or that may be added are a mixture of classes and will be administered as such. The administrative guidelines aim to "protect and enhance" the quality of various areas of a river.

Soon after a study of a potential wild

river is launched, public meetings are held to acquaint people near the river with the purposes, methods, and timing of the study. Later the study team reports its findings at further meetings and asks for reactions of local landowners and other interested individuals and groups. The results of these meetings are included in a final report, which becomes the basis for a proposal to Congress. The study teams work closely with state and local governments and any Federal agency that might have an interest in the river. The final report must give information concerning the effect on uses of the river and of nearby land if the river wins wild river status, along with current land ownership and use along the river, details of administering the river, and the estimated cost to the Federal government. Before the final proposal goes to Congress and the President, it is formally reviewed by the governor (or governors, if more than one state is involved) and by the concerned Federal agencies, whose comments become part of the report. It is then up to Congress to decide whether a proposed river is added to the national system.

Although this is the basic process for adding rivers to the national system, under certain conditions the Secretary of the Interior can add rivers on application from state governors. To qualify for the protection and prestige of the national system, a river must be protected by a state administrative program, at no cost to the Federal government. The Wild and Scenic Rivers Act emphasizes that states should undertake as much as possible of the job of preserving and administering wild and scenic rivers. The Bureau of Outdoor Recreation and the Forest Service assist states in establishing state and local wild, scenic, and recreational river areas and provide technical assistance and advice both to states and to private organizations. Although Federal money is not available for administering a river area once it is set up, funds are available from the Land and Water Conservation Fund on a fifty-fifty matching basis to plan, acquire, and develop state and local wild, scenic, and recreational rivers. Over a million dollars from the Land and Water Conservation Fund helped the State of Maine set up the Allagash Wilderness Waterway. Similarly, the Upper Wolf River in Wisconsin has been preserved as one of that state's wild and scenic rivers through the use of $537,586 in Federal funds.

The Wild and Scenic Rivers Act named the Allagash and Upper Wolf Rivers for potential inclusion in the national system, pursuant to state action. In July, 1970, at the request of the governor of Maine, the Secretary of the Interior added the Allagash, making it the ninth component of the national system.

The Wild and Scenic Rivers Act of 1968 is a complex but workable way of saving the best of the nation's rivers. It is complemented by the National Environmental Policy Act of 1969, which requires all Federal agencies to prepare detailed statements about the environmental impact of their projects, along with alternatives to them. As a result of this law, the future of the Corps of Engineers' projected Tocks Island Dam on the Delaware River was placed in doubt. Twice the Corps's statements on environmental impact were found inadequate by the President's Council on

Thousands of deer and elk find refuge deep in Hells Canyon
when severe winter weather makes survival impossible on
the surrounding mountains.

*The Snake River in Hells Canyon ranks close to the Colorado
in the Grand Canyon as a spectacular river-valley landscape.*

Environmental Quality. And in March, 1971, the United States District Court in Arkansas blocked, for a time at least, a Corps dam on the Cossatot River that was two thirds complete, Judge G. Thomas Eisele writing in his opinion that "the degree of the completion of the work should not inhibit the objective and thorough evaluation of the environmental impact of the project as required by the National Environmental Policy Act."

Judge Eisele's landmark decision goes on to declare that environmental impact statements should contain facts and viewpoints of concerned public or private organizations and private citizens, even if the responsible agency finds no merit in them whatsoever. "The record should be complete. . . . The most glaring deficiency in this respect," said the Judge, "is the failure to set forth and fully describe the alternative of leaving the Cossatot alone."

Conservationists can take heart in these decisions, and in the growing popularity and political strength of environmental causes. In 1963 only one state had an official system of free-flowing streams. That number has now been expanded to eighteen states, and many others have either authorized studies of possible state systems or are considering legislation to do so. So far the states have named more than six hundred rivers and streams as components of, or potential additions to, state systems.

The best of this nation's rivers and streams can be saved. The opportunity is there, and so is the challenge. At the 1971 Wilderness Conference, Jules Tileston of the Bureau of Outdoor Recreation suggested ways in which conservationists could assist the Wild and Scenic Rivers program. Reminding his audience that their help was needed to protect free-flowing streams not yet covered by the Act, he urged them to be alert, informed, and active in supporting or opposing specific proposals. He cited the 1970 report on the Tuolumne River prepared by the Sierra Club's Northern California Conservation Committee as "a concise document clearly setting out the conflicts and values involved . . . that . . . was largely responsible for the inclusion of the Tuolumne on the 5(d) list" and urged that similar nongovernmental studies on free-flowing river areas be made.

There is much to do on the state and local level. Most state wild and scenic river systems are in the fledgling or study stage, and some states have yet to take any action. The action will come when political leaders sense the need for it. In drafting legislation, the states should have no compunction about exceeding Federal standards. For example, the Federal law protects just a quarter mile of land along the banks of a river. Although Federal officials consider this sufficient to protect the rivers and allow public use, some rivers or parts of rivers, especially those in open country, may need much more protection. The Allagash Wilderness Waterway might be considered a model in this regard. No timber cutting or other disturbance is allowed within four hundred feet, and in some places eight hundred feet, of the river shore. The actual boundaries of the waterway, however, extend one mile from each shore. Though logging is permitted within these boundaries, it is carefully regulated by the state.

The boundaries of the Allagash Wilderness Waterway extend
a mile from each shore of the river.

On the national level, there are several ways in which people can act to save wild rivers. Some of the twenty-seven rivers being studied seemed destined for development when they were temporarily protected by the Wild and Scenic Rivers Act; controversy still surrounds them and only a great deal of public support, communicated to the U.S. Congress, will ensure their full protection. Other rivers may be rejected for nonpolitical reasons; Federal study teams have already concluded that some are just not scenic enough to merit inclusion in the national system, and one has been turned down because of a pollution problem that seems insoluble with present technology. However, this needn't exclude such streams from inclusion in river systems within their states.

In recent years, more than a dozen bills concerning wild rivers have been submitted to Congress. Some seek to bypass the complicated procedures of the Act and add rivers directly to the list now given full protection. Others are aimed at moving some of the rivers from the 5(d) list to the list that is scheduled for study (and protected for several years in the process). Still other legislation seeks to add further rivers to the 5(d) list. If legislation like this is ever to succeed, there must be greater support from anglers, hunters, whitewater enthusiasts, backpackers, and other concerned people who value these rivers.

Some bills would bypass the national system entirely. A bill creating the Buffalo National River in Arkansas seemed likely to become law in 1972; the legislation was similar to that which established the Ozark National Scenic Riverways in 1964. In recent years a variety of bills have been introduced to protect the Snake River in Hells Canyon —some proposed a seven-year moratorium on dams, and one sought to establish a Hells Canyon-Snake National River. Conservationists agree that some way must be found to preserve the Middle Snake; the victories so far in protecting wild rivers, and the victories to come, will be considerably diminished if this extraordinary river is lost.

The vigilance of conservationists should not end when a river is included in the national or state systems. Rivers in the national system will be administered by Federal agencies, state agencies, or a combination of both. The Act and its guidelines go into some detail on the administration of protected rivers:

Each component of the national wild and scenic rivers system shall be administered in such manner as to protect and enhance the values which caused it to be included in said system without, insofar as is consistent therewith, limiting other uses that do not substantially interfere with public use and enjoyment of these values. In such administration primary emphasis shall be given to protecting its esthetic, scenic, historic, archeologic, and scientific features.

The mandate seems clear, yet "protect and enhance" can be interpreted in different ways. For example, the guidelines for managing a wild river area include protection from fire, insects, and disease. This would protect and enhance the beauty of an area but might also make the area less natural, less like a truly wild ecosystem. An enhancement not mentioned in the guidelines is the re-establishment of flora

Deer, elk, and moose populations are usually healthiest and in proper balance with their food supply when culled by predators. These large herbivores would benefit if wolves and mountain lions were reestablished in certain wild river valleys.

and fauna that were once part of the area. In some wild river areas, the stocking of otters, cougars, and wolves would seem a natural enhancement of a wild river valley, providing there was sufficient wild land nearby to support the more terrestrial species. Controversy may arise over matters like these and over the conflicting goals of the Act—protecting and enhancing while allowing use and enjoyment. National parks have the same problems; it appears that "use and enjoyment" may have to be limited in order to protect parks from being trampled and developed to oblivion. Eventually, use of some popular components of the national wild river system may have to be limited for the same reason. Those who value wild rivers should familiarize them-selves with the Wild and Scenic Rivers Act, and its guidelines, so that they can help guard the protected rivers.

We have the means today of saving the finest of our undammed, unpolluted, and untamed rivers, and dedicating them to the purpose of great adventure for our minds and bodies. As long as these rivers run wild and free, there will be whitewater to test our muscles and skill, and trout to test our patience. There will be beautiful and complex places which man, for all his cleverness, cannot duplicate. There will be mayflies and otters obeying ancient laws which we have only begun to understand.

There will be all this and more, if only we have the good sense and love to keep these rivers flowing, full of life and mystery.

The Snake River within Grand Teton National Park has been
preserved for the enjoyment of future generations. The
fate of many other wild rivers rests with citizens
who know their value and can teach others.

Rivers Selected for 5(d) Status under the Wild and Scenic Rivers Act

Alaska

BIRCH CREEK—Segment from North Fork bridge at milepost 94 of the Steese Highway to highway bridge at milepost 147 of the Steese Highway.

CHATANIKA—Segment from the head of McManus Creek to the bridge at milepost 11 of the Elliott Highway.

CHITINA—The entire river.

DELTA—Segment from Round Tangle Lake at milepost 21 of the Denali Highway to the Delta's confluence with Phelan Creek at milepost 212.5 of the Richardson Highway.

FORTYMILE—Entire river with major tributaries within Alaska.

GULKANA—Entire main stem and its Middle and West Forks between the lower end of Paxson Lake and the town of Gulkana.

California

KERN—Segment from source of Kernville at Lake Isabella.

KLAMATH—Segment from Iron Gate Dam to mouth.

RUSSIAN—Segment from Ukiah to mouth.

SACRAMENTO—Segments from source to Shasta Lake and from Keswick Reservoir to Sacramento.

SMITH—Entire main stem, North Fork to Diamond Creek, Middle Fork to Griffen Creek, entire South Fork.

TUOLUMNE—Segment from Hetch-Hetchy Dam to New Don Pedro Reservoir.

Florida

WACISSA—The entire river.

Idaho

HENRYS FORK—Segment from Big Springs to confluence with Warm River.

SNAKE (MIDDLE)—Segment from Hells Canyon Dam to Lewiston, Idaho, including tributary Imnaha (also in Oregon and Washington).

Iowa

WAPSIPINICON—Segment within Linn, Bremer, Black Hawk, Buchanan, Jones and Clinton counties.

Louisiana

TANGIPAHOA—The entire river (also in Mississippi).

Maryland

POCOMOKE—The entire river.

Michigan

AUSABLE—Segment from Mio Pond to Alcona Hydro Plant.

MANISTEE—Segment from Hinton Creek to Hodenpyl Dam, and from Tippy Dam Pond to Manistee Lake, including tributary, Pine River from Stronach Dam at Tippy Dam Pond to Edgette's Bridge.

Minnesota

BIG FORK—Segment from confluence with Popple River to confluence with Rainy River.

Mississippi

TANGIPAHOA—The entire river (also in Louisiana).

Missouri

NORTH FORK WHITE RIVER—Segment from State Highway 76 to Lake Norfolk.

Montana

BLACKFOOT—Segment from Landers Fork to Milltown Dam.

MADISON—Segment from Earthquake Lake to Ennis Lake.

YELLOWSTONE—Segment from Yellowstone National Park boundary to Pompey's Pillar.

Nebraska

NIOBRARA—Segment from Antelope Creek to vicinity of Sparks.

New Jersey

MULLICA—Entire river including tributaries, Wading River and Bass River.

New York

BEAVERKILL—The entire river.

HUDSON—Segment from source to Luzerne, including tributaries.

North Dakota

LITTLE MISSOURI—Segment from Marmarth, N.D., to Lake Sakakawea.

Oregon

DESCHUTES—Segment from Pelton Reregulating Dam to confluence with Columbia.

GRAND RONDE—Segment from Rondowa to confluence with Snake River, with its tributaries, the Wenaha to Milk Creek on the South Fork of the Wenaha; the Wallowa to the Minam River; and the Minam in its entirety (also in Washington).

JOHN DAY—Segment from mouth to confluence with North Fork, North Fork from John Day at Kimberly to junction with Baldy Creek; Granite Creek to its junction with Clear Creek.

SNAKE (MIDDLE)—Segment from Hells Canyon Dam to Lewiston, Idaho, including tributary Imnaha (also in Idaho and Washington).

Texas

GUADALUPE—From source to Canyon Reservoir.

Utah

ESCALANTE—Source to Lake Powell.

Virginia

RAPPAHANNOCK—Segment from tidewater to Remington, including tributary Rapidan to community of Rapidan.

SHENANDOAH—The entire river (also in West Virginia).

Washington

COLUMBIA—Segment from Priest River Dam to McNary Pool.

GRAND RONDE—Segment from Rondowa to confluence with Snake River, with its tributaries, the Wenaha to Milk Creek on the South Fork of the Wenaha; the Wallowa to the Minam River; and the Minam in its entirety (also in Oregon).

SNAKE (MIDDLE)—Segment from Hells Canyon Dam to Lewiston, Idaho, including tributary Imnaha (also in Idaho and Oregon).

WENATCHEE—Entire river, including Lake Wenatchee and its tributaries, the Chiwawa and White rivers.

West Virginia

CACAPON—The entire river.

SHENANDOAH—The entire river (also in Virginia).

Wisconsin

FLAMBEAU (SOUTH FORK)—Segment from Round Lake to confluence with main stem.

PINE—Segment from source to confluence with Menominee River, including tributary, Popple River.

WOLF (UPPER)—Segment which flows through Langlade County.

Wyoming

GREEN (UPPER)—Source to Horse Creek.

GROS VENTRE—Entire river.

SNAKE—Segments from source in Yellowstone National Park to Jackson Lake, and from Jackson Lake to Palisades Reservoir.

WIND—Segment from source to Boysen Reservoir.

Environmental Organizations

The organizations listed below are actively engaged in the struggle to protect wild lands and the rivers and other streams that flow through them. Much of the success so far in preserving wild and scenic rivers can be credited to these groups. The Committee of Two Million is devoted to saving the few remaining undammed rivers in California; several of the other organizations have local, state, or regional chapters. For a more complete listing of environmental organizations, see the *Conservation Directory* published annually by the National Wildlife Federation.

Committee of Two Million
760 Market Street
San Francisco, California 94102

Friends of the Earth
451 Pacific Avenue
San Francisco, California 94133

The Izaak Walton League of America
719 13th Street, N.W.
Washington, D.C. 20005

National Audubon Society
950 Third Avenue
New York, New York 10022

National Parks and Conservation Association
1701 18th Street, N.W.
Washington, D.C. 20009

National Wildlife Federation
1412 16th Street, N.W.
Washington, D.C. 20036

The Nature Conservancy
1800 North Kent Street
Arlington, Virginia 22209

Sierra Club
1050 Mills Tower
San Francisco, California 94104

Trout Unlimited
4260 East Evans Avenue
Denver, Colorado 80222

The Wilderness Society
729 15th Street, N.W.
Washington, D.C. 20005

Index

Page numbers in italics indicate photographs.